Gravity: A Very Short Introduction

VERY SHORT INTRODUCTIONS are for anyone wanting a stimulating and accessible way into a new subject. They are written by experts, and have been translated into more than 45 different languages.

The series began in 1995, and now covers a wide variety of topics in every discipline. The VSI library now contains over 500 volumes—a Very Short Introduction to everything from Psychology and Philosophy of Science to American History and Relativity—and continues to grow in every subject area.

Very Short Introductions available now:

ACCOUNTING Christopher Nobes
ADOLESCENCE Peter K. Smith
ADVERTISING Winston Fletcher
AFRICAN AMERICAN RELIGION Eddie S. Glaude Jr
AFRICAN HISTORY John Parker and Richard Rathbone
AFRICAN RELIGIONS Jacob K. Olupona
AGEING Nancy A. Pachana
AGNOSTICISM Robin Le Poidevin
AGRICULTURE Paul Brassley and Richard Soffe
ALEXANDER THE GREAT Hugh Bowden
ALGEBRA Peter M. Higgins
AMERICAN HISTORY Paul S. Boyer
AMERICAN IMMIGRATION David A. Gerber
AMERICAN LEGAL HISTORY G. Edward White
AMERICAN POLITICAL HISTORY Donald Critchlow
AMERICAN POLITICAL PARTIES AND ELECTIONS L. Sandy Maisel
AMERICAN POLITICS Richard M. Valelly
THE AMERICAN PRESIDENCY Charles O. Jones
THE AMERICAN REVOLUTION Robert J. Allison
AMERICAN SLAVERY Heather Andrea Williams
THE AMERICAN WEST Stephen Aron
AMERICAN WOMEN'S HISTORY Susan Ware
ANAESTHESIA Aidan O'Donnell
ANARCHISM Colin Ward
ANCIENT ASSYRIA Karen Radner
ANCIENT EGYPT Ian Shaw
ANCIENT EGYPTIAN ART AND ARCHITECTURE Christina Riggs
ANCIENT GREECE Paul Cartledge
THE ANCIENT NEAR EAST Amanda H. Podany
ANCIENT PHILOSOPHY Julia Annas
ANCIENT WARFARE Harry Sidebottom
ANGELS David Albert Jones
ANGLICANISM Mark Chapman
THE ANGLO-SAXON AGE John Blair
THE ANIMAL KINGDOM Peter Holland
ANIMAL RIGHTS David DeGrazia
THE ANTARCTIC Klaus Dodds
ANTISEMITISM Steven Beller
ANXIETY Daniel Freeman and Jason Freeman
THE APOCRYPHAL GOSPELS Paul Foster
ARCHAEOLOGY Paul Bahn
ARCHITECTURE Andrew Ballantyne
ARISTOCRACY William Doyle
ARISTOTLE Jonathan Barnes
ART HISTORY Dana Arnold
ART THEORY Cynthia Freeland
ASIAN AMERICAN HISTORY Madeline Y. Hsu
ASTROBIOLOGY David C. Catling

ASTROPHYSICS James Binney
ATHEISM Julian Baggini
AUGUSTINE Henry Chadwick
AUSTRALIA Kenneth Morgan
AUTISM Uta Frith
THE AVANT GARDE David Cottington
THE AZTECS Davíd Carrasco
BABYLONIA Trevor Bryce
BACTERIA Sebastian G. B. Amyes
BANKING John Goddard and John O. S. Wilson
BARTHES Jonathan Culler
THE BEATS David Sterritt
BEAUTY Roger Scruton
BEHAVIOURAL ECONOMICS Michelle Baddeley
BESTSELLERS John Sutherland
THE BIBLE John Riches
BIBLICAL ARCHAEOLOGY Eric H. Cline
BIOGRAPHY Hermione Lee
BLACK HOLES Katherine Blundell
BLOOD Chris Cooper
THE BLUES Elijah Wald
THE BODY Chris Shilling
THE BOOK OF MORMON Terryl Givens
BORDERS Alexander C. Diener and Joshua Hagen
THE BRAIN Michael O'Shea
THE BRICS Andrew F. Cooper
THE BRITISH CONSTITUTION Martin Loughlin
THE BRITISH EMPIRE Ashley Jackson
BRITISH POLITICS Anthony Wright
BUDDHA Michael Carrithers
BUDDHISM Damien Keown
BUDDHIST ETHICS Damien Keown
BYZANTIUM Peter Sarris
CALVINISM Jon Balserak
CANCER Nicholas James
CAPITALISM James Fulcher
CATHOLICISM Gerald O'Collins
CAUSATION Stephen Mumford and Rani Lill Anjum
THE CELL Terence Allen and Graham Cowling
THE CELTS Barry Cunliffe
CHAOS Leonard Smith
CHEMISTRY Peter Atkins
CHILD PSYCHOLOGY Usha Goswami
CHILDREN'S LITERATURE Kimberley Reynolds
CHINESE LITERATURE Sabina Knight
CHOICE THEORY Michael Allingham
CHRISTIAN ART Beth Williamson
CHRISTIAN ETHICS D. Stephen Long
CHRISTIANITY Linda Woodhead
CITIZENSHIP Richard Bellamy
CIVIL ENGINEERING David Muir Wood
CLASSICAL LITERATURE William Allan
CLASSICAL MYTHOLOGY Helen Morales
CLASSICS Mary Beard and John Henderson
CLAUSEWITZ Michael Howard
CLIMATE Mark Maslin
CLIMATE CHANGE Mark Maslin
COGNITIVE NEUROSCIENCE Richard Passingham
THE COLD WAR Robert McMahon
COLONIAL AMERICA Alan Taylor
COLONIAL LATIN AMERICAN LITERATURE Rolena Adorno
COMBINATORICS Robin Wilson
COMEDY Matthew Bevis
COMMUNISM Leslie Holmes
COMPLEXITY John H. Holland
THE COMPUTER Darrel Ince
COMPUTER SCIENCE Subrata Dasgupta
CONFUCIANISM Daniel K. Gardner
THE CONQUISTADORS Matthew Restall and Felipe Fernández-Armesto
CONSCIENCE Paul Strohm
CONSCIOUSNESS Susan Blackmore
CONTEMPORARY ART Julian Stallabrass
CONTEMPORARY FICTION Robert Eaglestone
CONTINENTAL PHILOSOPHY Simon Critchley
COPERNICUS Owen Gingerich
CORAL REEFS Charles Sheppard
CORPORATE SOCIAL RESPONSIBILITY Jeremy Moon
CORRUPTION Leslie Holmes
COSMOLOGY Peter Coles
CRIME FICTION Richard Bradford

CRIMINAL JUSTICE Julian V. Roberts
CRITICAL THEORY
Stephen Eric Bronner
THE CRUSADES Christopher Tyerman
CRYPTOGRAPHY Fred Piper and
Sean Murphy
CRYSTALLOGRAPHY A. M. Glazer
THE CULTURAL
REVOLUTION Richard Curt Kraus
DADA AND SURREALISM
David Hopkins
DANTE Peter Hainsworth and
David Robey
DARWIN Jonathan Howard
THE DEAD SEA SCROLLS Timothy Lim
DECOLONIZATION Dane Kennedy
DEMOCRACY Bernard Crick
DEPRESSION Jan Scott and
Mary Jane Tacchi
DERRIDA Simon Glendinning
DESCARTES Tom Sorell
DESERTS Nick Middleton
DESIGN John Heskett
DEVELOPMENTAL BIOLOGY
Lewis Wolpert
THE DEVIL Darren Oldridge
DIASPORA Kevin Kenny
DICTIONARIES Lynda Mugglestone
DINOSAURS David Norman
DIPLOMACY Joseph M. Siracusa
DOCUMENTARY FILM
Patricia Aufderheide
DREAMING J. Allan Hobson
DRUGS Les Iversen
DRUIDS Barry Cunliffe
EARLY MUSIC Thomas Forrest Kelly
THE EARTH Martin Redfern
EARTH SYSTEM SCIENCE Tim Lenton
ECONOMICS Partha Dasgupta
EDUCATION Gary Thomas
EGYPTIAN MYTH Geraldine Pinch
EIGHTEENTH-CENTURY BRITAIN
Paul Langford
THE ELEMENTS Philip Ball
EMOTION Dylan Evans
EMPIRE Stephen Howe
ENGELS Terrell Carver
ENGINEERING David Blockley
ENGLISH LITERATURE Jonathan Bate
THE ENLIGHTENMENT
John Robertson
ENTREPRENEURSHIP Paul Westhead
and Mike Wright
ENVIRONMENTAL ECONOMICS
Stephen Smith
ENVIRONMENTAL POLITICS
Andrew Dobson
EPICUREANISM Catherine Wilson
EPIDEMIOLOGY Rodolfo Saracci
ETHICS Simon Blackburn
ETHNOMUSICOLOGY Timothy Rice
THE ETRUSCANS Christopher Smith
EUGENICS Philippa Levine
THE EUROPEAN UNION John Pinder
and Simon Usherwood
EVOLUTION
Brian and Deborah Charlesworth
EXISTENTIALISM Thomas Flynn
EXPLORATION Stewart A. Weaver
THE EYE Michael Land
FAMILY LAW Jonathan Herring
FASCISM Kevin Passmore
FASHION Rebecca Arnold
FEMINISM Margaret Walters
FILM Michael Wood
FILM MUSIC Kathryn Kalinak
THE FIRST WORLD WAR
Michael Howard
FOLK MUSIC Mark Slobin
FOOD John Krebs
FORENSIC PSYCHOLOGY
David Canter
FORENSIC SCIENCE Jim Fraser
FORESTS Jaboury Ghazoul
FOSSILS Keith Thomson
FOUCAULT Gary Gutting
THE FOUNDING FATHERS
R. B. Bernstein
FRACTALS Kenneth Falconer
FREE SPEECH Nigel Warburton
FREE WILL Thomas Pink
FRENCH LITERATURE John D. Lyons
THE FRENCH REVOLUTION
William Doyle
FREUD Anthony Storr
FUNDAMENTALISM Malise Ruthven
FUNGI Nicholas P. Money
GALAXIES John Gribbin
GALILEO Stillman Drake
GAME THEORY Ken Binmore
GANDHI Bhikhu Parekh
GENES Jonathan Slack

GENIUS Andrew Robinson
GEOGRAPHY John Matthews and David Herbert
GEOPOLITICS Klaus Dodds
GERMAN LITERATURE Nicholas Boyle
GERMAN PHILOSOPHY Andrew Bowie
GLOBAL CATASTROPHES Bill McGuire
GLOBAL ECONOMIC HISTORY Robert C. Allen
GLOBALIZATION Manfred Steger
GOD John Bowker
GOETHE Ritchie Robertson
THE GOTHIC Nick Groom
GOVERNANCE Mark Bevir
GRAVITY Timothy Clifton
THE GREAT DEPRESSION AND THE NEW DEAL Eric Rauchway
HABERMAS James Gordon Finlayson
HAPPINESS Daniel M. Haybron
THE HARLEM RENAISSANCE Cheryl A. Wall
THE HEBREW BIBLE AS LITERATURE Tod Linafelt
HEGEL Peter Singer
HEIDEGGER Michael Inwood
HERMENEUTICS Jens Zimmermann
HERODOTUS Jennifer T. Roberts
HIEROGLYPHS Penelope Wilson
HINDUISM Kim Knott
HISTORY John H. Arnold
THE HISTORY OF ASTRONOMY Michael Hoskin
THE HISTORY OF CHEMISTRY William H. Brock
THE HISTORY OF LIFE Michael Benton
THE HISTORY OF MATHEMATICS Jacqueline Stedall
THE HISTORY OF MEDICINE William Bynum
THE HISTORY OF TIME Leofranc Holford-Strevens
HIV AND AIDS Alan Whiteside
HOBBES Richard Tuck
HOLLYWOOD Peter Decherney
HOME Michael Allen Fox
HORMONES Martin Luck
HUMAN ANATOMY Leslie Klenerman
HUMAN EVOLUTION Bernard Wood
HUMAN RIGHTS Andrew Clapham
HUMANISM Stephen Law
HUME A. J. Ayer
HUMOUR Noël Carroll
THE ICE AGE Jamie Woodward
IDEOLOGY Michael Freeden
INDIAN CINEMA Ashish Rajadhyaksha
INDIAN PHILOSOPHY Sue Hamilton
THE INDUSTRIAL REVOLUTION Robert C. Allen
INFECTIOUS DISEASE Marta L. Wayne and Benjamin M. Bolker
INFORMATION Luciano Floridi
INNOVATION Mark Dodgson and David Gann
INTELLIGENCE Ian J. Deary
INTELLECTUAL PROPERTY Siva Vaidhyanathan
INTERNATIONAL LAW Vaughan Lowe
INTERNATIONAL MIGRATION Khalid Koser
INTERNATIONAL RELATIONS Paul Wilkinson
INTERNATIONAL SECURITY Christopher S. Browning
IRAN Ali M. Ansari
ISLAM Malise Ruthven
ISLAMIC HISTORY Adam Silverstein
ISOTOPES Rob Ellam
ITALIAN LITERATURE Peter Hainsworth and David Robey
JESUS Richard Bauckham
JOURNALISM Ian Hargreaves
JUDAISM Norman Solomon
JUNG Anthony Stevens
KABBALAH Joseph Dan
KAFKA Ritchie Robertson
KANT Roger Scruton
KEYNES Robert Skidelsky
KIERKEGAARD Patrick Gardiner
KNOWLEDGE Jennifer Nagel
THE KORAN Michael Cook
LANDSCAPE ARCHITECTURE Ian H. Thompson
LANDSCAPES AND GEOMORPHOLOGY Andrew Goudie and Heather Viles
LANGUAGES Stephen R. Anderson
LATE ANTIQUITY Gillian Clark
LAW Raymond Wacks
THE LAWS OF THERMODYNAMICS Peter Atkins
LEADERSHIP Keith Grint

LEARNING Mark Haselgrove
LEIBNIZ Maria Rosa Antognazza
LIBERALISM Michael Freeden
LIGHT Ian Walmsley
LINCOLN Allen C. Guelzo
LINGUISTICS Peter Matthews
LITERARY THEORY Jonathan Culler
LOCKE John Dunn
LOGIC Graham Priest
LOVE Ronald de Sousa
MACHIAVELLI Quentin Skinner
MADNESS Andrew Scull
MAGIC Owen Davies
MAGNA CARTA Nicholas Vincent
MAGNETISM Stephen Blundell
MALTHUS Donald Winch
MANAGEMENT John Hendry
MAO Delia Davin
MARINE BIOLOGY Philip V. Mladenov
THE MARQUIS DE SADE John Phillips
MARTIN LUTHER Scott H. Hendrix
MARTYRDOM Jolyon Mitchell
MARX Peter Singer
MATERIALS Christopher Hall
MATHEMATICS Timothy Gowers
THE MEANING OF LIFE Terry Eagleton
MEASUREMENT David Hand
MEDICAL ETHICS Tony Hope
MEDICAL LAW Charles Foster
MEDIEVAL BRITAIN John Gillingham and Ralph A. Griffiths
MEDIEVAL LITERATURE Elaine Treharne
MEDIEVAL PHILOSOPHY John Marenbon
MEMORY Jonathan K. Foster
METAPHYSICS Stephen Mumford
THE MEXICAN REVOLUTION Alan Knight
MICHAEL FARADAY Frank A. J. L. James
MICROBIOLOGY Nicholas P. Money
MICROECONOMICS Avinash Dixit
MICROSCOPY Terence Allen
THE MIDDLE AGES Miri Rubin
MILITARY JUSTICE Eugene R. Fidell
MINERALS David Vaughan
MODERN ART David Cottington
MODERN CHINA Rana Mitter
MODERN DRAMA Kirsten E. Shepherd-Barr
MODERN FRANCE Vanessa R. Schwartz
MODERN IRELAND Senia Pašeta
MODERN ITALY Anna Cento Bull
MODERN JAPAN Christopher Goto-Jones
MODERN LATIN AMERICAN LITERATURE Roberto González Echevarría
MODERN WAR Richard English
MODERNISM Christopher Butler
MOLECULAR BIOLOGY Aysha Divan and Janice A. Royds
MOLECULES Philip Ball
THE MONGOLS Morris Rossabi
MOONS David A. Rothery
MORMONISM Richard Lyman Bushman
MOUNTAINS Martin F. Price
MUHAMMAD Jonathan A. C. Brown
MULTICULTURALISM Ali Rattansi
MUSIC Nicholas Cook
MYTH Robert A. Segal
THE NAPOLEONIC WARS Mike Rapport
NATIONALISM Steven Grosby
NELSON MANDELA Elleke Boehmer
NEOLIBERALISM Manfred Steger and Ravi Roy
NETWORKS Guido Caldarelli and Michele Catanzaro
THE NEW TESTAMENT Luke Timothy Johnson
THE NEW TESTAMENT AS LITERATURE Kyle Keefer
NEWTON Robert Iliffe
NIETZSCHE Michael Tanner
NINETEENTH-CENTURY BRITAIN Christopher Harvie and H. C. G. Matthew
THE NORMAN CONQUEST George Garnett
NORTH AMERICAN INDIANS Theda Perdue and Michael D. Green
NORTHERN IRELAND Marc Mulholland
NOTHING Frank Close
NUCLEAR PHYSICS Frank Close
NUCLEAR POWER Maxwell Irvine
NUCLEAR WEAPONS Joseph M. Siracusa
NUMBERS Peter M. Higgins
NUTRITION David A. Bender

OBJECTIVITY Stephen Gaukroger
THE OLD TESTAMENT
Michael D. Coogan
THE ORCHESTRA D. Kern Holoman
ORGANIZATIONS Mary Jo Hatch
PANDEMICS Christian W. McMillen
PAGANISM Owen Davies
THE PALESTINIAN-ISRAELI CONFLICT Martin Bunton
PARTICLE PHYSICS Frank Close
PAUL E. P. Sanders
PEACE Oliver P. Richmond
PENTECOSTALISM William K. Kay
THE PERIODIC TABLE Eric R. Scerri
PHILOSOPHY Edward Craig
PHILOSOPHY IN THE ISLAMIC WORLD Peter Adamson
PHILOSOPHY OF LAW
Raymond Wacks
PHILOSOPHY OF SCIENCE
Samir Okasha
PHOTOGRAPHY Steve Edwards
PHYSICAL CHEMISTRY Peter Atkins
PILGRIMAGE Ian Reader
PLAGUE Paul Slack
PLANETS David A. Rothery
PLANTS Timothy Walker
PLATE TECTONICS Peter Molnar
PLATO Julia Annas
POLITICAL PHILOSOPHY
David Miller
POLITICS Kenneth Minogue
POPULISM Cas Mudde and Cristóbal Rovira Kaltwasser
POSTCOLONIALISM Robert Young
POSTMODERNISM
Christopher Butler
POSTSTRUCTURALISM
Catherine Belsey
PREHISTORY Chris Gosden
PRESOCRATIC PHILOSOPHY
Catherine Osborne
PRIVACY Raymond Wacks
PROBABILITY John Haigh
PROGRESSIVISM Walter Nugent
PROTESTANTISM Mark A. Noll
PSYCHIATRY Tom Burns
PSYCHOANALYSIS Daniel Pick
PSYCHOLOGY Gillian Butler and Freda McManus
PSYCHOTHERAPY Tom Burns and Eva Burns-Lundgren
PUBLIC ADMINISTRATION
Stella Z. Theodoulou and Ravi K. Roy
PUBLIC HEALTH Virginia Berridge
PURITANISM Francis J. Bremer
THE QUAKERS Pink Dandelion
QUANTUM THEORY
John Polkinghorne
RACISM Ali Rattansi
RADIOACTIVITY Claudio Tuniz
RASTAFARI Ennis B. Edmonds
THE REAGAN REVOLUTION
Gil Troy
REALITY Jan Westerhoff
THE REFORMATION Peter Marshall
RELATIVITY Russell Stannard
RELIGION IN AMERICA Timothy Beal
THE RENAISSANCE Jerry Brotton
RENAISSANCE ART
Geraldine A. Johnson
REVOLUTIONS Jack A. Goldstone
RHETORIC Richard Toye
RISK Baruch Fischhoff and John Kadvany
RITUAL Barry Stephenson
RIVERS Nick Middleton
ROBOTICS Alan Winfield
ROCKS Jan Zalasiewicz
ROMAN BRITAIN Peter Salway
THE ROMAN EMPIRE
Christopher Kelly
THE ROMAN REPUBLIC
David M. Gwynn
ROMANTICISM Michael Ferber
ROUSSEAU Robert Wokler
RUSSELL A. C. Grayling
RUSSIAN HISTORY Geoffrey Hosking
RUSSIAN LITERATURE Catriona Kelly
THE RUSSIAN REVOLUTION
S. A. Smith
SAVANNAS Peter A. Furley
SCHIZOPHRENIA Chris Frith and Eve Johnstone
SCHOPENHAUER Christopher Janaway
SCIENCE AND RELIGION
Thomas Dixon
SCIENCE FICTION David Seed
THE SCIENTIFIC REVOLUTION
Lawrence M. Principe
SCOTLAND Rab Houston
SEXUALITY Véronique Mottier
SHAKESPEARE'S COMEDIES
Bart van Es
SIKHISM Eleanor Nesbitt

THE SILK ROAD James A. Millward
SLANG Jonathon Green
SLEEP Steven W. Lockley and Russell G. Foster
SOCIAL AND CULTURAL ANTHROPOLOGY John Monaghan and Peter Just
SOCIAL PSYCHOLOGY Richard J. Crisp
SOCIAL WORK Sally Holland and Jonathan Scourfield
SOCIALISM Michael Newman
SOCIOLINGUISTICS John Edwards
SOCIOLOGY Steve Bruce
SOCRATES C. C. W. Taylor
SOUND Mike Goldsmith
THE SOVIET UNION Stephen Lovell
THE SPANISH CIVIL WAR Helen Graham
SPANISH LITERATURE Jo Labanyi
SPINOZA Roger Scruton
SPIRITUALITY Philip Sheldrake
SPORT Mike Cronin
STARS Andrew King
STATISTICS David J. Hand
STEM CELLS Jonathan Slack
STRUCTURAL ENGINEERING David Blockley
STUART BRITAIN John Morrill
SUPERCONDUCTIVITY Stephen Blundell
SYMMETRY Ian Stewart
TAXATION Stephen Smith
TEETH Peter S. Ungar
TELESCOPES Geoff Cottrell
TERRORISM Charles Townshend
THEATRE Marvin Carlson
THEOLOGY David F. Ford
THOMAS AQUINAS Fergus Kerr
THOUGHT Tim Bayne
TIBETAN BUDDHISM Matthew T. Kapstein
TOCQUEVILLE Harvey C. Mansfield
TRAGEDY Adrian Poole
TRANSLATION Matthew Reynolds
THE TROJAN WAR Eric H. Cline
TRUST Katherine Hawley
THE TUDORS John Guy
TWENTIETH-CENTURY BRITAIN Kenneth O. Morgan
THE UNITED NATIONS Jussi M. Hanhimäki
THE U.S. CONGRESS Donald A. Ritchie
THE U.S. SUPREME COURT Linda Greenhouse
UTOPIANISM Lyman Tower Sargent
THE VIKINGS Julian Richards
VIRUSES Dorothy H. Crawford
VOLTAIRE Nicholas Cronk
WAR AND TECHNOLOGY Alex Roland
WATER John Finney
WEATHER Storm Dunlop
THE WELFARE STATE David Garland
WILLIAM SHAKESPEARE Stanley Wells
WITCHCRAFT Malcolm Gaskill
WITTGENSTEIN A. C. Grayling
WORK Stephen Fineman
WORLD MUSIC Philip Bohlman
THE WORLD TRADE ORGANIZATION Amrita Narlikar
WORLD WAR II Gerhard L. Weinberg
WRITING AND SCRIPT Andrew Robinson
ZIONISM Michael Stanislawski

Available soon:

MILITARY STRATEGY Antulio J. Echevarria
ANIMAL BEHAVIOUR Tristram D. Wyatt
NAVIGATION Jim Bennett
THE HABSBURG EMPIRE Martyn Rady
INFINITY Ian Stewart

For more information visit our website

www.oup.com/vsi/

Timothy Clifton

GRAVITY

A Very Short Introduction

Great Clarendon Street, Oxford, OX2 6DP,
United Kingdom

Oxford University Press is a department of the University of Oxford.
It furthers the University's objective of excellence in research, scholarship,
and education by publishing worldwide. Oxford is a registered trade mark of
Oxford University Press in the UK and in certain other countries

First edition published in 2017

Published in the United States of America by Oxford University Press
198 Madison Avenue, New York, NY 10016, United States of America

British Library Cataloguing in Publication Data
Data available

Library of Congress Control Number: 2016952558

ISBN 978–0–19–872914–3

Printed and bound by CPI Group (UK) Ltd, Croydon, CR0 4YY

Contents

Preface

This is a book about gravity. Gravity is what causes massive objects to fall towards each other. It's gravity that causes apples to fall from trees, and it was gravity that caused the Earth to form. Gravity is the most familiar of all the fundamental forces of nature, yet the true way in which gravity works is far from obvious. In reality, the phenomenon we refer to as 'gravity' is deeply tied up with the nature of space and time. This means that a modern understanding of gravity doesn't just tell us how objects in the Universe move, it also allows us to understand the behaviour of the very space and time that make up the fabric of the Universe itself.

The purpose of this book is to give the reader a very brief introduction to various different aspects of gravity. We start by looking at the way in which the theory of gravity developed historically, before moving on to an outline of how it is understood by scientists today. We will then consider the consequences of gravitational physics on the Earth, in the Solar System, and in the Universe as a whole. The final chapter describes some of the frontiers of current research in theoretical gravitational physics.

By the end, I intend to have conveyed to the reader not only what gravity is, but how the study of gravity has led scientists to reach extraordinary conclusions about the nature of space, time, and the Universe in which we live.

List of illustrations

Chapter 1
From Newton to Einstein

Gravity is by far the weakest of the four fundamental forces that exist in nature—the others being the electromagnetic force, and the strong and weak nuclear forces. Yet over large distances it is gravity that dominates. This is because gravity is only ever attractive and because it can never be screened. So while most large objects are electrically neutral, they can never be gravitationally neutral. The gravitational force between objects with mass always acts to pull those objects together, and always increases as they become more massive.

It is thanks to geniuses like Newton and Einstein that we understand gravity at all, yet it is gravity, more than any other force, that continues to present scientists with the most enigmatic of puzzles. To understand why, let's start at the beginning, and consider the historical development of gravity.

The pre-history of gravity

It seems safe to assume that mankind has always known that when we drop an object it will fall downwards. In this sense, we've always been aware of the existence of gravity. It was the cause of this motion that appears to have been the focus of many early thinkers.

Aristotle, who's *Physics* played a dominant role in European science until as late as the 17th century, based his explanation of gravity on the idea that objects should move towards their natural place in the Universe. Where this place was, he reasoned, should be determined by the composition of the object in question. More specifically, it should depend on how much it contained of each of the four elements: Earth, Water, Air, and Fire.

Aristotle argued that objects composed predominantly from Earth and Water should move towards the centre of the Universe. To him, the centre of the Universe was the place beneath his feet. Objects made from Earth, and that are thrown into the air, must therefore move down towards the ground. Water, he reasoned, is lighter than Earth, as can be verified by dumping some soil into a half-full glass of it. All Water must therefore come to rest on top of the Earth. Similarly, Air is lighter than Water, as bubbles rise in water. The natural place for Air is therefore above Water, and the natural place for Fire is above the Air.

This framework provides a logical sort of order to the world that was observed, in terms of what were thought to be the basic constituents of matter at the time. It even allows one to make some statements about the speed at which objects should fall. Aristotle argued that the velocity of a falling object should be in proportion to its weight, and in inverse proportion to the density of the medium through which it travels. That is, Aristotle thought that an object that weighs 2kg should fall twice as fast as one that weighs 1kg.

Sadly, Aristotle's theory cannot be correct. We now know that there is no centre to the Universe for objects to move towards. We can also demonstrate, by direct experimentation, that the rate at which an object accelerates due to gravity is *not* in proportion to its mass. In fact, it can be shown that all objects fall at the same rate. This discovery was one of the milestones in the modern understanding of gravity, and so it bears some further explanation.

The fact that all objects accelerate under gravity at the same rate is not an obvious one. In fact, if I were to drop a feather from my left hand, and a lump of iron from my right, I should not expect them to hit the floor at the same time. The lump of iron will get there first. So what is meant by the statement 'all objects accelerate under gravity at the same rate'? To understand this phrase, we have to think about all of the forces that act upon these objects.

When I let go of the feather it is acted on by the force of gravity, but it is also acted on by other forces. As it starts to drop, there is resistance to its motion from the air around it. This slows the feather more than it does the heavier lump of iron. Any slight gust of air will also have a large effect on the feather while perturbing the motion of the lump of metal by only a tiny amount. So, the meaning of the phrase 'all objects accelerate under gravity at the same rate' is not a statement about the motion of objects in our immediate environment. Rather, it is a statement about what should happen to an object if it were to fall under the influence of gravity alone. That is, if all other interactions are suppressed, then all objects should fall at the same rate.

Galileo is widely credited with having shown the truth of this proposition. In 1638, he supposedly dropped cannonballs with different masses off the top of the leaning tower of Pisa. The cannonballs were found to fall at the same rate, independent of their composition. More recently, and perhaps even more dramatically, the same result was demonstrated by the Apollo astronaut David Scott. Scott dropped a feather and a hammer while standing on the surface of the Moon. There is no air on the Moon to slow the motion of the feather, and so both objects landed at his feet at the same time (see Figure 1). Today we refer to this phenomenon as the *Universality of Free Fall*. It is a key ingredient in both Newton's and Einstein's theories of gravity, as we will see later.

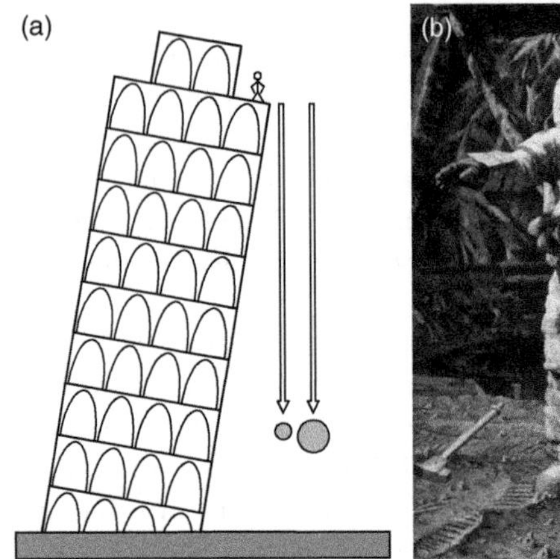

1. (a) Galileo dropping cannonballs off the leaning tower of Pisa; and (b) a painting of David Scott dropping a hammer and a feather on the surface of the moon.

Newton's theory of gravity

Sir Isaac Newton's theory of gravity and motion was first published in his *Principia Mathematica* in 1687, and it changed the world forever. This was the first genuinely scientific theory of how gravity worked. Unlike Aristotle, Newton didn't try and seek an explanation for gravity. Instead he quantified its effects, and in the process deduced physical laws that accurately described the motion of not only objects on Earth, but also the motion of the Earth itself, as well as all other bodies in the Solar System.

Newton's achievement was truly a work of genius. He created new branches of mathematics, and for the first time demonstrated that the laws of physics that apply to us, on Earth, are the same as those that apply to the heavenly bodies. All the complexity of motion, which Aristotle had tried to explain, was boiled down to a handful of simple laws. Newton's theory was glorious, and went unchallenged for over 200 years. In one book, Newton revolutionized the worlds of science, industry, and warfare, and provided the framework within which many of us still think today.

The basic ingredients of Newton's theory are the existence of absolute space and time as the stage on which all motion occurs, and the existence of a universal gravitational force that acts instantaneously between every pair of massive bodies in the Universe. And that is all.

Space to Newton was, as it appears to most of us in our day-to-day experience, simply the eternal and unchanging arena within which all matter exists. I can position object X at one point in space, and work out the distance to any other object, Y, by simply using a measuring tape held out in a straight line. In Newton's mechanics there is no ambiguity in this process. Objects X and Y may move in space, but the space itself is fixed and forever unchanged.

Likewise, Newton's conception of time followed the everyday intuition that most of us grow up with. Instants of time unfold in Newton's theory, one after the other. Objects can change their position during an interval of time, but time itself is universal and the same for everyone. In Newton's theory all true clocks measure time in the same way, just as all true measuring tapes measure the same distance between any two given objects.

Now, according to Newton, all bodies move at a constant rate, unless acted on by an external force (this was a departure from Aristotle's physics). If a force should act upon an object, then the effect of that force is to cause the object to accelerate. More force means more acceleration, and, if an object has more mass, then it needs more force to reach the same acceleration. Within this framework, gravity is simply an external force acting on all massive bodies, pulling them together.

Newton deduced that his gravitational force must be proportional to the mass of each of the objects it acts between, and inversely proportional to the square of the distance between them. That is, the force of gravity between two massive bodies must be given by an equation of the form:

$$F \propto \frac{Mm}{r^2},$$

where M and m are the masses of the two objects in question, and r is the distance between them. This simple equation, together with Newton's laws of motion, is sufficient to get a good approximation to the motion of the vast majority of astrophysical bodies as well as all bodies that exist on Earth.

That all bodies fall under Newton's gravity at the same rate, and hence obey Galileo's observation, can be seen from the fact that in Newton's mechanics a given force causes a heavy body to accelerate more slowly than a light one. Consider this together with the fact that, according to Newton, the gravitational force must increase with mass. In Newton's theory these two things happens at exactly the required rate to cancel each other out. A body being acted on by Newton's gravity, and obeying Newton's laws of motion, must therefore accelerate at a fixed rate, independent of its mass. That this happens is no accident: the Universality of Free Fall is built into Newton's theory from the start.

The first great success of Newton's theory was that it could be used to derive the laws of motion for the planets. These laws had been deduced empirically by Johannes Kepler earlier in the 17th century, using cutting-edge astronomical data. Kepler's laws stated that:

- The orbit of a planet traces out an ellipse, and the Sun is at one of the foci of this ellipse.
- If a line is drawn between a planet and the Sun, then, as the planet orbits the Sun, that line will sweep out an equal area of space in any two intervals of equal time.

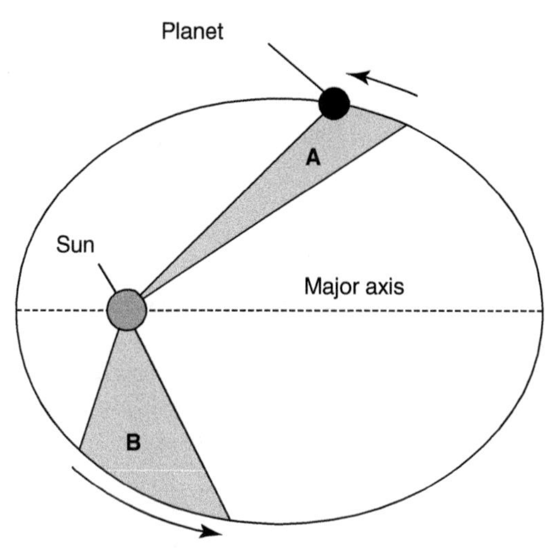

2. An illustration of a planetary orbit. The shaded regions, marked A and B, are of equal area, if they are swept out by the orbiting planet in equal time.

- The square of the time taken for a planet to complete one orbit is proportional to the cube of the distance between the two furthest points on that orbit (the major axis of the ellipse).

The quantities described in these three laws are all illustrated in Figure 2 (with the exception of the orbital period). The good thing about Kepler's laws was that they seemed to apply to all known planets, even though at the time they had no known basis in any physical theory. They were simply seen to be a good fit to the data.

Newton knew about Kepler's laws, and included in his *Principia Mathematica* an explanation of how they could be derived from his laws of motion and from his universal theory of gravitation. This derivation was one of the great success stories of physics. Empirical laws, derived from cutting-edge observations of the planets, were explained for the first time in terms of simple mathematical equations. Newton had shown that the same laws that describe the motion of a cannonball fired on the surface of the Earth could also be used to describe the motions of the planets themselves. This was the beginning of much of physics as we know it today.

Einstein's theory of gravity

More than 200 years after its publication, Newton's theory of gravity was supplanted by Albert Einstein's. If Newton's theory was simple and useful, then Einstein's was beautiful and truly universal. Einstein didn't just change the equations involved in Newton's theory, he pulled up the very foundations upon which it rested. Einstein changed everything.

As with much of the progress that happens in physics, Einstein's theory was motivated primarily by the inconsistency of existing ideas. Newton had given us a theory of how gravity and motion operate. But Newton's ideas were not compatible with the theory of light that James Clerk Maxwell had developed in the middle of the 19th century. Maxwell's new theory stated that everyone in the Universe should measure the speed of light to have the same value: just under 300 million metres per second. This might not sound terribly profound, until you take a few moments to consider what it means.

The point is that, according to Newton's mechanics, if I fire a bullet forwards at 1,000 miles per hour while seated on a train moving at 100 miles per hour, then an observer on the side of the track will see that bullet moving at 1,100 miles per hour. In mathematical terms we say that the velocity of the bullet and the velocity of the train add linearly. Now consider that I turn on a torch, while still seated facing forwards on the same train. From my seated position I will see the light from the torch propagate forwards through the train carriage at the speed of light (i.e. at about 300 million metres per second). Consider again the observer watching me from the side of the train track. If you listened to Newton, you might expect this person to see the light travel at the 300 million metres per second, plus 100 miles per hour (the speed of the train). But according to Maxwell this is not what happens. Maxwell says that the person at the side of the track sees the light propagate at the *same* speed as the person on the train. That is, Maxwell's equations imply that velocities do *not* add linearly.

The contradiction just described is a profound one. If we are unable to agree on how to add velocities, then we are unable to use physics to calculate the motion of objects at all. It was impossible for both Newton and Maxwell to be correct. At least one of them must be wrong. A lesser scientist might have tried to re-write either Newton's or Maxwell's theories, but this was not what Einstein did. Einstein treated both Newton's and Maxwell's work with the utmost respect. He recognized their great strengths, and worked to solve the contradiction in a truly ingenious way.

Einstein hypothesized that if the speed of light was the same for everyone, then time and space cannot be universal concepts. Instead, he reasoned, each observer must have their own personal concept of time, and their own personal concept of space. According to Einstein's new theory, a clock carried by a person on the train is seen, by a person standing beside the track, to tick slower than a clock they carry themselves. Likewise, the person on the train sees the clock of the person standing beside the track to tick slower than his or her own.

This result initially sounds odd, but that's only because we've been programmed from a young age to think of time as universal. What Einstein showed us is that our childhood understanding of time is mistaken. Time is not a universal concept, unfolding at the same rate for everyone. Time is a personal thing, and depends on our relative motion, with respect to others. Likewise, space is not the fixed backdrop that we think it is. What we think of as distances, and the lengths of objects, are actually dependent on how we are moving.

These are startling ideas. At first they can seem unsettling, as if the crutches we've used to understand the world have suddenly been kicked away. But we need not despair. There is a concept involving space and time that survives in Einstein's theory, and that maintains an observer-independent reality. This is what is known as *space-time*. Instead of Newton's concepts of universal time and universal space, what we are left with is a larger structure that encompasses them both. A person or object, like you or me, follows a line through this structure, known as our *world-line*. Our personal time is measured along our world-line, and while my world-line might be different to yours, they both exist in the same space-time (see Figure 3).

So, it is the promotion of space and time to space-time that allows us to make Newton's mechanics consistent with Maxwell's. This realization was one of Einstein's early contributions to science, and it is the backbone of what is now known as the *Special Theory*

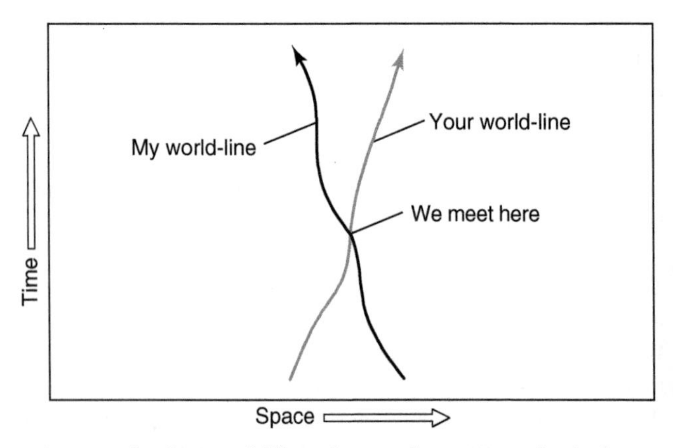

3. An example of two world-lines, for two observers, moving in the same space-time. The observers meet where the lines cross.

of Relativity. It has a wide variety of profound consequences, many of which have been confirmed experimentally. The most famous of these is probably the equation:

$$E = m\,c^2,$$

which tells us that mass and energy are intricately connected (a fact that became devastatingly apparent with the advent of nuclear weapons). Other consequences are the longer lifetime that unstable particles appear to have when they move quickly, and the fact that nothing can ever move faster than light.

It is this last result, together with the new concept of space-time, which led Einstein to his theory of gravity. Again, the impetus for this development was an apparent contradiction. And again, it was Newton's theory that was causing problems. This time, however, it seemed that there was a contradiction with Einstein's own work. This was because Newton's gravity acted between objects instantaneously. That is, if the Sun were to suddenly explode, according to Newton we should feel the gravitational effects of this at the very same moment it happens. But Einstein now knew that this was impossible. First, he had found out that nothing could move faster than light. Second, he had shown that there was no such thing as universal time, so the idea of two things happening simultaneously in two different places made no sense at all (if they happened simultaneously for one observer, they would not be simultaneous for any other who was in a different state of motion). So, once again, something was wrong and needed fixing.

Einstein's solution to this problem was even more amazing. He hypothesized that gravity, instead of being a force that simply pulled things through space, was the result of the curvature of space-time. The fact that massive objects were drawn towards each other was then, according to Einstein, just a result of those objects following the shortest paths they could in the curved space-time in which they existed. The idea was that mass and energy caused space-time to curve, and that this curvature caused the paths of the objects that move through space to appear to bend towards each other. The beauty of this idea is that we now no longer need to include gravity as an extra force that exists in the Universe. In this new picture, the only thing responsible for the attraction of massive bodies is space-time itself (which has to be there anyway). This is the fundamental idea behind the *General Theory of Relativity*.

Even more impressive is that Einstein's idea explained Galileo's result that all objects fall at the same rate. Recall that in Newton's theory this result wasn't really explained at all. It was simply taken as a fact, and a law of gravity was devised that was compatible with it. Einstein went one better. Now, in Einstein's theory, there is no external force called gravity; the motion of every object is just a result of the curvature of space-time. But all objects are moving in the same space-time, so all objects must follow the same paths. In other words, all objects must fall at the same rate, just as Galileo had observed.

These ideas can sometimes be confusing, so let's think about an example. Imagine the paths of two objects that have no forces acting on them. In a flat space the paths of these objects are straight lines, as shown in Figure 4.

If the space is curved, however, then this is no longer true. Consider the simplest curved space: the surface of a sphere. The shortest path between any two points on the surface of a sphere is called a *great circle* (the equator is an example of a great circle—on the globe). If two objects follow two different great circles, on the same sphere, then they will initially move away from each other, before finally coming together again, as shown in Figure 5.

This is how Einstein envisaged gravity working. He imagined that it was curvature that was responsible for the paths of objects meeting, and not anything external that pulls them left or right as

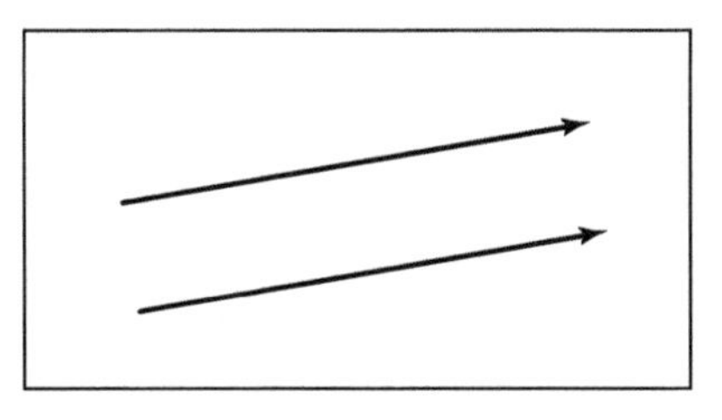

4. Two straight lines, indicating the paths that two particles might follow if they travelled through a flat space, without any external forces acting upon them.

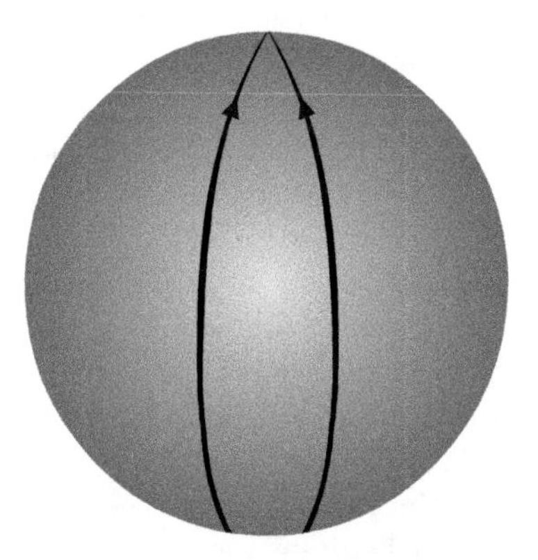

5. Two great circles, indicating the paths that particles might travel in a spherically curved space, if no external forces act upon them. The lines are no longer parallel forever, but meet at a point.

they move. The curvature of space-time is usually more irregular than the surface of a sphere, but the basic idea is the same. As far as the consequences for gravity go, the largest effects of Einstein's new theory were found to look very similar to the law that Newton had prescribed 200 years earlier. The difference is that this law now comes with a new understanding of space and time. It also comes with the prediction of a variety of smaller, more subtle, effects (to be discussed in the following chapters).

Think now about what all of this means in a practical situation. Consider a skydiver jumping out of an airplane. The skydiver falls freely, up to the effects of air resistance. According to Einstein, the skydiver's path is the straightest line possible through the curved space-time around the Earth. From the skydiver's perspective this seems quite natural. Except for the air rushing past her, the skydiver feels no perturbing forces at all. In fact, if it weren't for the air resistance, she would experience weightlessness in the same way that an astronaut does in orbit. The only reason we think the skydiver is accelerating is because we are used to using the surface of the Earth as our frame of reference. If we free ourselves from this convention, then we have no reason to think the skydiver is accelerating at all.

Now consider yourself on the ground, looking up at the falling daredevil. Normally, your intuitive description of your own motion would be that you are stationary. But again this is only because of our slavish regard to the Earth as the arbiter of what is at rest and what is moving. Free yourself from this prison, and you realize that you are, in fact, accelerating. You feel a force on the soles of your feet that pushes you upwards, in the same way that you would if you were in a lift that accelerated upwards very quickly. In Einstein's picture there is no difference between your experience standing on Earth and your experience in the lift. In both situations you are accelerating upwards. In the latter situation it is the lift that is responsible for your acceleration. In the former, it is the fact that the Earth is solid that pushes you upwards through space-time, knocking you off your free-fall trajectory. That the surface of the Earth can accelerate upwards at every point on its surface, and remain a solid object, is because it exists in a curved space-time and not in a flat space.

With this change in perspective the true nature of gravity becomes apparent. The freely falling skydiver is brought to Earth because the space-time through which she falls is curved. It is not an external force that tugs her downwards, but her own natural motion through a curved space. On the other hand, as a person standing on the ground, the pressure you feel on the soles of your feet is due to the rigidity of the Earth pushing you upwards. Again, there is no external force pulling you to Earth. It is only the electrostatic forces in the rocks below your feet that keep the ground rigid, and that prevents you from taking what would otherwise be your natural motion (which would also be free fall).

So, if we free ourselves from defining our motion with respect to the surface of the Earth we realize that the skydiver is *not* accelerating, while the person who stands on the surface of the Earth *is* accelerating. Just the opposite of what we usually think. Going back to Galileo's experiment on the leaning tower of Pisa, we can now see why he observed all of his cannonballs to fall at the same rate. It wasn't really the cannonballs that were accelerating away from Galileo at all, it was Galileo that was accelerating away from the cannonballs! If I leave a number of objects at rest at some position in space and accelerate away from them, I shouldn't be at all surprised that the distance between me and each of these objects increases at the same rate. So it is with Galileo and his cannonballs.

For some, the beauty in this description is obvious. For others it is the fact that it is possible to probe Einstein's theory experimentally that is most compelling. These experiments range from looking for small perturbations to the orbits of the planets, to the bending of light around the Sun, and many, many more besides. We will explore these exciting phenomena in the chapters that follow. But bear in mind throughout: it is the curvature of space-time that is responsible for it all.

Chapter 2
Gravity in the Solar System

The Solar System, including the Earth, is our most immediate laboratory for observing the consequences of gravity. The gravitational field in the Solar System is dominated by the Sun, which is far more massive than any of the planets.

In orbits that are relatively close to the Sun are the four smallest planets: Mercury, Venus, Earth, and Mars. Further out, there are four much larger planets: Jupiter, Saturn, Uranus, and Neptune. The Solar System contains a number of other objects, such as comets, asteroids, moons, and man-made spacecraft. By observing the motion of these objects, and in some cases by interacting with them, we can learn a great deal about the behaviour of gravity.

The study of experimental and observational gravity in the Solar System took off in earnest during the latter half of the 20th century. While astronomers had been tracking the motion of the planets for centuries, the development of new technologies and methods in the 20th century allowed observations and experiments to be carried out in ways that had never previously been possible. In order to try and present the results of this work in some kind of sensible order, I will classify them into experiments that probe the foundational assumptions of gravity theories; experiments that probe Newton's law; and experiments that probe the subtle effects that result from Einstein's theory.

Testing foundational assumptions

There are a number of foundational assumptions that go into the modern theory of gravity. These include the fact that the rest mass of an object should be independent of its position, and independent of its motion with respect to other bodies. Other assumptions are the idea that the speed of light should be the same in every direction, and that all objects should fall at the same rate (in the absence of non-gravitational forces). Over the course of the past century all these ideas have been tested to extremely high accuracy. I'll outline some of the best of these tests here, before we move on to thinking about experiments that test the particulars of Newton's and Einstein's theories of gravity.

Let's start off by thinking about mass. Recall that mass is the quantity that tells us how much force we need to apply to an object in order to make it accelerate by a fixed amount. It's thought to be a property of the object itself. This is different to weight, which is the name of the downwards force that an object exerts on your hand when you hold it, and which would be different if you held the same object while standing on a different planet. It is mass that appears in Newton's law of gravity, and it is mass that was shown to be equivalent to energy in Einstein's famous formula ($E = mc^2$). Because these two equations are so central to gravity, we need to have some idea of whether or not the mass of an object really is independent of its position and motion in a gravitational field. This can only be done experimentally.

Oddly enough, the best test of whether mass depends on position is to look at how light changes colour as it travels through a gravitational field. The basic idea behind this test is that photons (particles of light) should lose energy as they escape from the gravitational fields of massive objects such as stars or planets. This is because it takes energy to pull something upwards through a gravitational field. So, just as you would have to use energy to run up a flight of stairs, it takes a photon some amount of energy to travel upwards from the surface of the Earth, or away from the surface of the Sun. A change in energy of a photon results in a change in its colour (its wavelength), and so a beam of light travelling through a gravitational field should be expected to have a different colour depending on how far it is above the source of that field. This is the reason why light detected far from the surface of a star is found to be slightly redder, or longer in wavelength, than it was when it was emitted. The effect is known as the *gravitational redshifting* of light.

So, how can the gravitational redshifting of light be expected to tell us anything about whether the mass of an object depends on its position in a gravitational field? To answer this, let's consider the most direct way of measuring the mass of an object in a gravitational field: lifting it upwards with a winch and recording the amount of energy it took to do so. This energy should be directly related to the mass of the object, so recording the energy required to lift the object between two different heights should tell us something very direct about its mass at, and between, those two positions. Unfortunately, it's difficult to accurately measure the amount of energy used by a winch, as they tend to be quite inefficient (they give out energy in noise and heat, and by stretching the rope and their own components). This is where the gravitational redshifting of light comes in. The frequency of light can be measured to very high precision, and the energy the photons lose by climbing out of the gravitational field is expected to be exactly the same as that which would be used to lift an object with a mass that corresponds to the same energy (as calculated using $E = mc^2$). If we can measure the redshifting of light, we therefore have a highly accurate proxy experiment, which is expected to convey exactly the same information as the winch experiment.

An experiment designed to measure the gravitational redshifting of light was performed for the first time by the scientists Robert

Pound and Glen Rebka in the early 1960s. They used a tower at the Jefferson Physical Laboratory at Harvard University, and looked at the redshifting of photons as they travelled up its height. They found that the light did indeed change colour as it travelled upwards, and by an amount that was entirely consistent with mass being independent of position. The only deviations possible would have to be smaller than the accuracy of the experiment itself, which was at the level of about 1 per cent. A similar experiment has since been carried out using the light emitted from the Sun, which also had results that were consistent with mass being independent of position, again at an accuracy of about 1 per cent.

More recently, atomic clocks have been used to probe the same effect. The idea behind these experiments is that beams of light are themselves, in some sense, like clocks. The colour of light is determined by the wavelength of the photons that make up the light, which is itself related to the frequency at which they oscillate. If we were to take these oscillations to be the basis of a clock, by saying that each oscillation is one unit of time, then we can see that the redshifting of light is equivalent to the appearance of clocks at different positions running at different rates. In fact, we don't even need to measure the frequency of light to perform this experiment, because if a clock based on the oscillations of a photon appears to run slow, then so must every other clock. All we have to do is put two clocks at two different heights and arrange for them to transmit the time they display via radio signals. The difference in the rate at which they appear to tick, according to the radio signals that are observed, is then entirely equivalent to the rate at which a light signal between them should be redshifted.

We can therefore take two highly accurate atomic clocks and put one on a rocket while leaving the other at our feet. If the clock in the rocket transmits its time via radio signals, then we can compare this signal to the time displayed by the clock we kept beside us. They should, in general, be different: an effect known as *gravitational time dilation*. This experiment was carried out for the first time by Robert Vessot and Martin Levine in 1976. They observed the time dilation effect directly, and found it to be consistent with mass being independent of position to an accuracy of about 1 part in 10,000 (about a hundred times more accurate than the Harvard experiment). This experiment therefore provides very strong evidence that mass is indeed independent of position in a gravitational field, to very high accuracy.

Moving on from position dependence, it is possible to construct experiments that test whether the speed of light, or the mass of an object, are dependent on direction. These experiments are historically very important, because before Einstein produced his theory of gravity it was widely believed that there was a substance called *the ether* that permeated the whole of space. The ether was invoked as the medium through which waves of light travelled, and was popular among physicists until as late as the 20th century. If the ether existed, then the speed of light that an observer measured should depend on their motion as they travelled through it. Einstein's theory is incompatible with the existence of an ether, as he had constructed it so that all observers measure the same speed of light, independent of their state of motion. Performing tests for the existence of the ether was therefore an important part of verifying his theory. The most famous of these tests was the Michelson–Morley experiment, carried out in 1887. It is a test to see if the speed of light depends on its direction of propagation.

The Michelson–Morley experiment used a device known as an *interferometer*, consisting of two arms, built at right-angles to each other. This is illustrated in Figure 6. Light from a laser is shone down each of the arms, and reflected back on itself by mirrors positioned at each of their ends. When the reflected light reached the intersection of the two arms it was allowed to interact. As light is known to have the properties of a wave, it would be possible to let the two beams of light interact to form a pattern

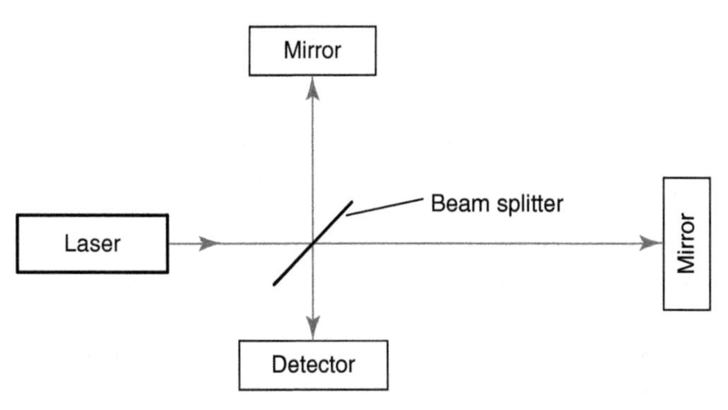

6. A schematic diagram of an interferometer. The beam splitter separates the laser light into two beams, each of which is reflected off a mirror. The light then re-traces its path, and interacts, before being directed into the detector.

(like two sets of waves interacting on the surface of a pond). The form of the pattern that is produced in the interferometer depends on the length of each of the arms, and the time it takes the laser light to travel along them. If the speed of light was different in different directions, then Michelson and Morley should have seen the consequences of this by using their apparatus.

Michelson and Morley's experiment had a null result. No difference was observed in the speed of light, as it travelled in the two different directions. This was unexpected by many scientists at the time, as the Earth was supposed to be in motion with respect to the ether. If light was a wave in the ether, then the speed of light should have been direction independent only in a laboratory that was stationary with respect to the ether. This wasn't thought to be the case for the Earth, which orbits the Sun at around 30,000 metres per second. The experiment performed by Michelson and Morley is therefore taken to be strong evidence against the existence of an ether, and in favour of a speed of light that is truly the same in every direction. This was very important for Einstein's theory.

Evidence for the direction independence of mass was given by the experiments that were independently performed by Vernon Hughes and Ronald Drever in the early 1960s. These investigators considered the electrons that exist within lithium atoms, and that orbit the nucleus at velocities of around a million metres per second. Now, the gravitational interaction between these electrons, and anything else in their immediate environment, is extremely small. This is because electrons have such tiny masses. Nevertheless, it is still possible to put extremely tight bounds on any possible direction dependence of their mass. This is because electrons give out photons when they change energy levels within atoms, and because these photons have a very specific set of frequencies known as *transition lines*. If the mass of the electrons depended on their direction of motion, then this would change the precise position of the transition lines that resulted from changes in energy levels. Careful studies, by both Hughes and Drever, found no evidence for the direction dependence of the mass of the electron, to extremely high accuracy.

Let's now return to the idea of the Universality of Free Fall. Remember that this is the name given to the idea that all objects fall at the same rate, as demonstrated by Galileo. But Galileo's experiment, while ground-breaking, was probably not very accurate (by modern standards). So, given the enormous significance of the Universality of Free Fall for both Newton and Einstein's theories, there has subsequently been a lot of effort to verify it to the highest possible level of accuracy. This has now been done in a number of different environments, including numerous laboratory experiments as well as space-based observations.

The most famous of the laboratory tests for the Universality of Free Fall was performed by Loránd Eötvös at the end of the 19th century. Eötvös used a device called a *torsion balance*, which consisted of a 40cm bar suspended by a fibre at its mid-point. This is illustrated in Figure 7. Two different objects were then hung

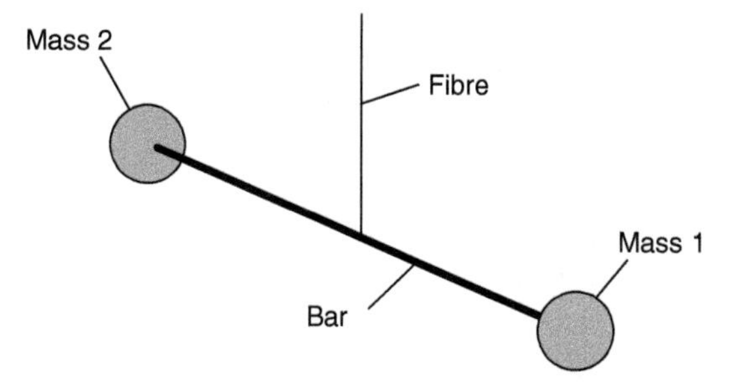

7. A schematic illustration of the torsion balance experiment, used by Eötvös. The two masses are made from different materials. If the masses were to fall at different rates, then the bar would rotate around the fibre.

from the two ends of the bar. These objects were chosen to be made of two very different materials. The idea was that if the two objects accelerated towards the Earth at different rates then the torsion balance would be forced to rotate. By measuring the motion of the beam it would then be possible to place limits on the difference between the gravitational force that acted on each of the objects and hence to put limits on the different rates at which they would fall if they were released.

Eötvös found that there was no difference in the rates at which any two objects would fall, to an accuracy of around one part in a billion. This, then, was a very precise version of Galileo's experiment. Since the 19th century the accuracy of this type of experiment has been improved still further, with groups at Princeton, the University of Washington, and Moscow University increasing its precision to better than one part in a trillion. This progress was made possible by the availability of technologies that have allowed the experiment to be performed in near vacuum conditions, as well as because the experimenters now make use of the gravitational field of the Sun in addition to that of the Earth. The current limiting factors in these experiments are the tiny rumblings from the Earth's shifting tectonic plates, together with the tiny gravitational fields from other nearby objects (including from the experimenters themselves!). Space-based experiments are being considered that would remove some of these difficulties and increase the accuracy of the experiments still further.

An alternative approach to the Eötvös experiments is to use the Earth and Moon themselves as the two objects in free fall, and see if a difference in their acceleration can be detected. This idea was made possible after 1969, when the astronauts from Apollo 11 left a reflector on the moon. By bouncing a laser beam off this reflector it is now possible to determine the distance between the Earth and the Moon to an accuracy of about 1cm. Careful analysis of the resultant data has allowed the Universality of Free Fall to be probed to the level of about one part in a billion. Although this accuracy does not exceed the lab-based results, it is a slightly different type of experiment, as the Earth and Moon have their own sizeable gravitational fields. This allows a 'strong' version of the same principle to be tested.

So, to summarize, there now exists good evidence, from a variety of different experiments, that the mass of an object is independent of both its position and the direction of its motion. There is also strong evidence that the speed of light is the same in every direction, and that all objects fall under gravity at the same rate. These are the foundational assumptions under which Newton's and Einstein's theories were constructed. Let's now move on to experiments that aim to investigate these theories themselves.

Experiments that probe Newton's law

Later on in this chapter we will consider the predictions of Einstein's theory, which is now universally accepted to supplant that of Newton. However, we know that Newton's inverse square law of gravity is a good enough approximation to Einstein's theory to describe a wide variety of both terrestrial and astronomical phenomena. It's therefore very much worthwhile testing Newton's law to see exactly how well it describes gravity. Here we will consider some of the most important tests that have been performed so far.

The first laboratory test of the inverse square law was performed by Henry Cavendish, at the end of the 18th century. Cavendish's experiment made use of a torsion balance, like the one used by Eötvös (illustrated in Figure 7). Unlike Eötvös, however, Cavendish positioned extra masses in order to make the torsion balance rotate. He put these masses close to the ones that were attached to the torsion bar. The gravitational force between the test objects hanging from the torsion balance, and the extra masses that Cavendish introduced, could then be inferred by the rate of rotation of the apparatus. The new masses pulled at the masses attached to the torsion bar through the gravitational force between them, and caused the bar to rotate. The results of Cavendish's experiment thus allowed for the gravitational force to be probed over a distance scale of only 23cm. He found that gravity operated on these scales in a way that was entirely consistent with Newton's inverse square law. Today, experimenters have tested these same ideas on much smaller distance scales.

The challenges involved in testing Newton's law of gravity in the laboratory arise principally due to the weakness of the gravitational force compared to the other forces of nature. This weakness means that even the smallest residual electric charges on a piece of experimental equipment can totally overwhelm the gravitational force, making it impossible to measure. All experimental equipment therefore needs to be prepared with the greatest of care, and the inevitable electric charges that sneak through have to be screened by introducing metal shields that reduce their influence. This makes the construction of laboratory experiments to test gravity extremely difficult, and explains why we have so far only probed gravity down to scales a little below 1mm (this can be compared to around a billionth of a billionth of a millimetre for the electric force).

There are three groups of experimenters that have led the way in laboratory tests of gravity in recent times: one at the University of Washington, another at the University of Colorado, and a third at Stanford University. The Washington group took a pendulum with ten holes cut into it and held it over a disk that contained another set of holes. The pendulum was found to twist due to the missing gravitational force from the missing mass in the holes, and this allowed gravity to be measured down to scales as small as a twentieth of a millimetre. The groups at Colorado and Stanford, on the other hand, measured the gravitational field of a vibrating mass, and used this information to probe gravity all the way down to 1/40 of a millimetre. The results of these laboratory experiments have so far all been entirely consistent with Newton's inverse square law of gravity, even on the very smallest of scales.

On larger distance scales, however, there are a variety of other tests that we must consider. To order the discussion of these tests sensibly, let's first of all consider distances from tens of metres, up to about to about a few kilometres. These scales may seem like they should be the easiest to probe as they are closest to the type of distances we consider in our everyday lives. They are, however, quite problematic.

One of the best constraints that has so far been achieved on gravity over these everyday distance scales is due to measurements of the gravitational force that an object experiences at various heights up a tall tower. Towards the end of the 1980s, a team of scientists performed this test using the 600-metre-high WTVD tower in Garner, North Carolina. The change in gravity expected from Newton's inverse square law at different points up the tower is easy to calculate, and the experimenters were able to measure it to good accuracy. At around the same time, other researchers were measuring the gravitational force due to the water in a reservoir filled to different levels. In essence, this allowed the water to be weighed and Newton's inverse square law of gravity to be tested. Further tests, a couple of years later, were performed by measuring the gravitational attraction of sea water at different depths in the ocean. These experiments all produced results that were consistent with Newton's inverse square law, and all at a level of accuracy of around 0.1 per cent.

On larger scales it starts to become possible to use astronomical data, which is much more accurate than trying to measure the gravitational field of the water in reservoirs and oceans. Distance scales of between a million and a few hundreds of millions of metres can be probed by studying the orbits of man-made satellites around the Earth, the orbit of the Moon, and the orbits of the planets around the Sun. The LAGEOS satellites, launched in 1976 and 1992, are particularly useful for experiments of this type—as are the orbits of the planets, which form closed ellipses (to first approximation) when the gravitational attraction is of the form given by Newton's inverse square law. Observations of all of these bodies, and objects, have shown consistency with Newton's law at accuracies of between one part in a million and one part in a billion.

We therefore have very good evidence that Newton's inverse square law is a good approximation to gravity over a wide range of distance scales. These scales range from a fraction of a millimetre, to hundreds of millions of metres. The accuracy of the experiments that probe gravity over these distances varies wildly, between about 1 part in 1,000 (for scales of tens of metres) to about one part in a billion (for distances that correspond to the orbits of the planets). This is a great success, but it's not the end of the story. Let's now go beyond Newton's law and start considering the new effects that arise from Einstein's theory.

Experiments that probe Einstein's theory

The experiments discussed earlier have all involved concepts that are familiar to most people who have studied physics at school: the constancy of mass; the Universality of Free Fall; and Newton's inverse square law of gravity. In this section, I will introduce experiments that probe the more unfamiliar ground of Einstein's theory of gravity. The specific effects that result from this theory are often small and difficult to detect experimentally. Nevertheless, they are very important, as they provide us with a window that can be used to view and understand gravity at a deeper level.

There are a large number of effects that result from Einstein's theory. Here I am going to limit myself to describing four of them. These are the anomalous orbit of the planet Mercury; the bending of starlight around the Sun; the time delay of radio signals as they pass by the Sun; and the behaviour of gyroscopes in orbit around the Earth. These are four of the most prominent relativistic gravitational effects that can be observed in the Solar System. Further effects, which become apparent in more extreme astrophysical environments, will be discussed in Chapter 3.

Let's start with the anomalous orbit of Mercury. Earlier on in this book, I stated that Newton's theory explained Kepler's observation that the planets follow elliptical orbits around the Sun. This is true for a single planet, but things start to get a bit more complicated when we consider the orbits of multiple planets simultaneously. This is because there exist gravitational forces between the planets themselves and not just between each planet and the Sun individually. These interplanetary gravitational forces are smaller but measurable, and have the effect of pulling the orbits of the planets off what might otherwise be perfect ellipses.

Physicists have known about the effects of interplanetary gravitational forces for a long time. These forces can be easily calculated within Newton's theory, and astronomers have measured their consequences on the orbits of the planets for centuries. In fact, the existence of the planet Neptune was deduced in the mid-19th century by carefully studying the orbit of Uranus (the next nearest planet to the Sun). The orbit of Uranus was perturbed a little from where astronomers expected it to be, and the perturbation could be explained if there was a slightly larger planet a little bit further out in the Solar System. Urbain Le Verrier and John Couch Adams both predicted the existence of this planet in 1845—and by 1846 it had been observed. This was clearly a momentous achievement.

It therefore wasn't too much of a shock when, in 1859, Le Verrier announced that the orbit of Mercury (the planet closest to the Sun) also seemed to deviate a little from the path that it was expected to take. No doubt buoyed by his prediction of the existence of Neptune, Le Verrier predicted that there must be another planet still closer to the Sun than Mercury. He even named it—*Vulcan.* This time, however, the discovery of the predicted planet never occurred. Despite much effort, no new object could be seen between Mercury and the Sun. Nevertheless, the anomalous orbit of Mercury persisted. It seemed as if its orbit was being dragged around the Sun by a gravitational force that had no obvious origin.

The mystery of the anomalous orbit of Mercury was solved in 1915, not by the discovery of any new massive bodies in the Solar System, but by Einstein's revolutionary new theory. You see, according to Einstein's new theory, the Newtonian gravitational force is only a rough approximation to the true nature of gravity. As well as the inverse square law, Einstein predicted that there should be new, smaller contributions to the gravitational force law. For an astrophysical system dominated by a large mass at the centre, like the Sun at the centre of the Solar System, Einstein calculated that the largest of these new forces should vary as the inverse of the distance cubed. Therefore, the closer one is to the Sun, the larger the contribution from this new force should be, relative to Newton's inverse square law.

Mercury was, and still is, the closest known planet to the Sun, and so Einstein's new gravitational force should be more influential on Mercury's orbit than on the orbits of any of the other planets.

Einstein calculated that Mercury's orbit should be dragged around the Sun by an extra 43 arcseconds per century (1 arcsecond is 1/360 of a degree). This is a tiny amount, but it is enough to be observed by astronomers. It is also consistent with the anomalous observations made by Le Verrier. So Einstein's theory of gravity had its first observational success as early as 1915, by explaining the anomalous orbit of the planet Mercury.

Modern measurements of the orbit of Mercury are much better than those that existed in the 19th century. We now know the orbits of all the planets to very high accuracy, which is essential for calculating the deviation of Mercury's orbit from an ellipse. The extra drag caused by Venus alone, for example, is more than six times that of the correction due to Einstein's gravity. Its position therefore needs to be known to very high accuracy. This isn't the largest source of uncertainty in modern observations though. That distinction goes to the uncertainty associated with the shape of the Sun itself. Any deviations from a perfectly spherical shape cause effects in the orbit of Mercury that could be confused with Einstein's new gravitational force. The shape of the Sun is hard to know exactly, so the best we can currently do is to say that the anomalous orbit of Mercury is consistent with Einstein's theory to an accuracy of about 1 part in 1,000.

The explanation of Mercury's orbit is impressive but probably doesn't count as a prediction, because the anomaly itself was known about long before Einstein was even born. What does count as a genuine prediction of Einstein's theory is the bending of starlight around the Sun by its gravitational field. It was previously uncertain whether or not light should be affected by gravity, because in Newton's theory the gravitational force is only between objects that have mass (which light does not). This ambiguity was removed in Einstein's theory, as light simply follows the shortest paths available in the curved space-time, just like everything else. Einstein therefore predicted that light should be bent by the gravitational fields of massive objects.

Einstein's calculations showed that the deflection of light would be greatest for beams that just skim the surface of massive objects. The most massive object in the Solar System is the Sun. But to see starlight that passes very close to the Sun we have to wait for a solar eclipse, otherwise the Sun's own light overwhelms that of the much fainter stars. The first good opportunity to test the idea of bending starlight came after the end of World War I, in 1919. The expedition to measure the positions of the stars nearby the Sun, and therefore to test Einstein's theory of gravity, was led by Sir Arthur Eddington.

Eddington's expedition went to the island of Príncipe, in Africa, where the solar eclipse that was going to happen that year would be total. He made careful measurements, using the best photographic plates that were available at the time. The conditions were less than ideal, but Eddington succeeded in measuring the positions of the stars during the eclipse. He found that they had indeed shifted, just as Einstein had predicted, because the trajectory of the light had been bent by the gravitational influence of the Sun. Eddington's results were sufficient to confirm Einstein's theory, but only within an accuracy of around 30 per cent.

Once again, modern observations have improved this result dramatically. This has been helped partly by the fact that there are, coincidentally, a number of very bright objects, known as *quasars*, which lie in just the right part of the sky to test Einstein's prediction. As these objects pass behind the Sun, the deflection of the light that they emit can be measured. Several million observations of these quasars have now been made using very large baseline *interferometers* (this is a type of telescope that mixes the signals from a number of different detectors to create a high-resolution image). The result of this work shows perfect consistency with Einstein's theory, at a level of accuracy of about 1 part in 10,000.

A more recent prediction from Einstein's theory is the time delay of radio signals that pass by massive objects. For some reason, it took until 1964 for scientists to notice that this effect is a necessary outcome of Einstein's gravity, but it has now been measured in a number of different situations. These have included the observation of radio signals that bounce off planets as they are about to pass behind the Sun, as well as by looking at the signals that are emitted from man-made probes as they do the same. The first of these methods has the benefit of the position of the planets being very well known, and their trajectories being easy to predict. This stability makes them a good target, but the imperfections in their shape can cause some problems in interpreting the reflected signals. Man-made probes, on the other hand, emit very easily predictable signals, but their trajectories can be somewhat less certain.

Radio targets that have been used to measure the time delay effect include the planets Mercury and Venus, and the space probes Mariners 6 and 7, Voyager 2, the Viking Mars Lander and Orbiters, and the Cassini probe. The most recent, and most accurate, of these observations was made using Cassini. The primary mission of this spacecraft was to observe Saturn, but its relevance for the study of gravity was perhaps best served in 2003, when it was announced it had confirmed the existence of the time delay effect with an accuracy of 1 part in 100,000. This was yet another spectacular confirmation of Einstein's theory, and one which was achieved at higher accuracy than any previous experiment. This was, in part, due to the observations of the radio signals being made at multiple frequencies, which allowed interference from the Sun's corona to be extracted.

Now let's move on to the final experiment I want to discuss in this section: the behaviour of *gyroscopes* in orbit around the Earth. A gyroscope is essentially a spinning top, whose central axis is allowed to point in any direction. According to Einstein's theory of gravity, there should be two new effects that can be observed when a gyroscope is put in orbit around the Earth. The first of these is a change in direction of the gyroscope's axis of rotation, as it orbits the Earth. This effect, known as *geodetic precession*, is due to the curvature of space-time around the Earth. The second effect is known as *frame-dragging*, and is due to the rotation of the Earth effectively pulling space around with it as it rotates. This is an entirely new type of gravitational interaction, and so it is of great interest to see if it can be confirmed experimentally.

Although the prediction of the frame-dragging effect was made only a few years after Einstein published his new theory of gravity, it took until the 1960s before the effect on gyroscopes in orbit was calculated, and the experimental observation of the effects themselves did not take place until the 21st century. The LAGEOS satellite network provided an observation of the frame-dragging effect by measuring the change in the orbit of its satellites as they went around the rotating Earth. The long-awaited gyroscope experiment was performed by a mission called *Gravity Probe B*, in 2011. The geodetic precession and frame-dragging effects were measured by this experiment with accuracies of about 0.3 per cent and 20 per cent, respectively. The accuracy of the corresponding results from the LAGEOS satellites are estimated at between 5 and 10 per cent. All results were once again found to be consistent with Einstein's theory.

So the overall picture we are left with is very encouraging for Einstein's theory of gravity. The foundational assumptions of this theory, such as the constancy of mass and the Universality of Free Fall, have been tested to extremely high accuracy. The inverse square law that formed the basis of Newton's theory, and which is a good first approximation to Einstein's theory, has been tested from the sub-millimetre scale all the way up to astrophysical scales. We are also now in possession of a number of accurate experimental results that probe the tiny, subtle effects that result from Einstein's theory specifically. This data allows us direct experimental insight into the relationship between matter and the curvature of space-time, and all of it is so far in good agreement with Einstein's predictions. This is a truly spectacular confirmation of a theory that was borne out of almost pure thought. Einstein wanted to make a theory of gravity that was compatible with a speed of light that was measured to be the same by every observer. He did so, and we have now seen the numerous consequences of his revolutionary new picture of the Universe. This isn't the end of the story though: Einstein's gravity has many more spectacular consequences, which we will discuss in the following chapters.

Chapter 3
Extrasolar tests of gravity

The Solar System allows us to investigate a number of different gravitational effects. Many of them can be measured to high accuracy, because we have easy access to the nearby planets and satellites. They are, however, quite weak gravitational fields. This is because all of the objects in the Solar System are, relatively speaking, rather slow moving and not very dense. If we set our sights a little further though, we can find objects that are much more extreme than anything we have available nearby.

Let's start by considering the life of a star. First-generation stars are thought to form when the clouds of hydrogen gas that emerged from the Big Bang collapse under their own gravitational field, and become hotter and denser. Eventually nuclear reactions start occurring, and the outward pressure from these reactions becomes strong enough to balance the inward attraction of gravity. This results in a star: a hot ball of collapsed gas undergoing violent nuclear reactions. This process of collapsing gas and nuclear reactions is, of course, also a rough description of what happens in our own Sun.

But this isn't the end of the story. A star such as the Sun can only ever have a finite lifetime. Eventually the hydrogen required for nuclear fusion will run out, and the star will have to start burning other materials. This makes it swell into a red giant. In turn, even these alternative fuels must eventually run out, and the gravitational collapse of the star will resume. What stops the collapse at this point depends on the size of the star. A small star will settle down to become a *white dwarf*. This is a state in which the quantum mechanical properties of the electron prevent it from becoming any smaller. At this point it's just not possible to fit any more electrons in the space that the star takes up.

If a star is a bit bigger, then instead of becoming a white dwarf it will end up as a *neutron star*. In a star of this type, the end of nuclear fusion leads to a collapse of its core. The star then collapses, which in turn causes a colossal explosion, known as a *supernova*. During this process the gravitational force is strong enough to force the electrons and protons to combine into neutrons. The electron pressure is then absent, and the star collapses until the neutrons are so dense that no more can fit into the same space. The end result is a star that has a density that's comparable to the nucleus of an atom. In a sense, a neutron star can be thought of as a giant atomic nucleus (but without any protons and without electrons orbiting it). Neutron stars are very small and very dense. They are denser than anything that exists in the Solar System, and tend to move at extremely high velocities.

A neutron star isn't the most extreme object that can result from a collapsing star though. That title goes to a type of object called a *black hole*. If a star is so large that even the pressure from neutrons can't support it, then it will collapse to a black hole. A black hole is one of the most extreme objects that can exist in nature. All that is really left after a collapse of this type is the gravitational field itself. A black hole consists of a region of space-time enclosed by a surface called an *event horizon*. The gravitational field of a black hole is so strong that anything that finds its way inside the event horizon can never escape. Even light. Hence the name.

In this section we are going to consider star systems that contain some of the objects just described. Astronomers have now discovered a large number of these systems, and observations of them have allowed us to explore gravity in ways that are simply impossible in our own Solar System. The extreme nature of these objects amplifies the effects of Einstein's theory, so that, even though they are very far away, they offer us a new and exciting window through which to see the effects of gravity.

The *Hulse–Taylor binary pulsar*

The *Hulse–Taylor binary pulsar*, or PSR B1913+16, is a star system that contains two neutron stars in orbit around each other. The remarkable thing about this system is that one of the neutron stars is what is known as a *pulsar* (an abbreviation of 'pulsating star'). These are stars that appear to emit regular pulses of radiation, when viewed from Earth. This radiation is the result of the strong magnetic fields that surround the star and that cause powerful beams to be projected outwards from its surface. Together with the rapid rotation of neutron stars, these beams appear as rapid flashes of radiation to distant astronomers, much like the signal from a lighthouse appears as flashes of light to nearby sailors.

Pulsars were discovered for the first time in 1967, by Jocelyn Bell Burnell and Antony Hewish. These astronomers saw the characteristic flashes of radiation that are now known to signal the existence of a pulsar, but at the time were quite unexpected. Indeed, it was originally thought that these flashes could be signals from another civilization. The astronomers even went as far as to call the source of the signal *LGM-1* (Little Green Men-1). Later, similar signals were discovered from other parts of the sky, and it was realized that the pulses were from a rapidly rotating neutron star. At present we know of the existence of thousands of pulsars. In the future we are likely to find many more.

The significance of the discovery of PSR B1913+16 by Joseph Taylor and Russell Hulse in 1974 was not that it was a pulsar, but that it was a pulsar in orbit around another neutron star. This information was obtained by noticing that the arrival times of the pulses varied slightly. That is, the arrival times of the pulses were sometimes three seconds earlier and sometimes three seconds later. The period over which this change happened was about seven hours and forty-five minutes. As the pulsar emitted seventeen pulses a second, it was possible to make a smooth chart that showed a clear pattern. The only explanation was that the pulsar was in orbit around another object, and that the radius of the orbit was about three light seconds (that is, the distance light travels in a period of three seconds, equal to about a million kilometres).

So the Hulse–Taylor pulsar was known to be part of a binary system, but it was still not possible to see the other object in the system. This meant it couldn't be a regular star, but it had to be something with a similar mass to a star. It was decided that the most plausible explanation was that the pulsar was part of a binary system, with the other body in the system being a neutron star that wasn't pulsating (or, at least, wasn't sending any pulses of radiation in our direction). Such a system is of particular interest for the study of gravity because the objects involved are so dense and are orbiting each other at such extremely high velocities. This makes the small effects that are predicted from Einstein's theory much more prominent. The fact that one of the neutron stars was emitting a signal that was as accurate as an atomic clock was the icing on the cake. Information could be extracted from this signal about the details of the gravitational interaction.

Data has been collected from the Hulse–Taylor binary pulsar since its discovery in 1974. This has been done primarily with the Arecibo telescope in Puerto Rico, which is a 305-metre-wide radio antenna (and which will be familiar to anyone who has seen the film *GoldenEye*). It is this large collection of data that makes this particular binary pulsar system special. Later, I will discuss some other binary pulsar systems that are now known to exist, but none of these have been observed for as long as the Hulse–Taylor pulsar. The large database that has been constructed for this system allows for some very precise tests of gravity to be performed.

Let's now consider the specifics of how information about gravity is encoded in the pulsar's signal. One way that this happens is through time delays and redshift effects that the signal experiences, as it travels through the gravitational field of the companion neutron star. You will recall that both of these effects have been measured in the Solar System. Now they can be measured in two distant neutron stars, which orbit each other at a distance that is similar to the radius of our own Sun. Another effect that will be familiar, and that is visible in the Hulse–Taylor binary system, is the precession of the orbit. Just as the orbit of Mercury precesses around the Sun so too the neutron stars in the Hulse–Taylor binary system precess around each other. To compare with similar effects in our Solar System, the orbit of the Hulse–Taylor pulsar precesses as much in a day as Mercury does in a century.

A final effect which it is possible to measure with the binary pulsar, and which is impossible to measure in the Solar System, is the change in period of an orbit due to the emission of gravitational waves. The existence of gravitational waves has not yet been discussed, but it is a definite prediction of Einstein's theory: there should exist ripples in space-time that can carry energy out of a system. This is a phenomenon that has no counterpart in Newton's theory, and is therefore of particular interest for testing Einstein's gravity. I will explain gravitational waves in more detail in Chapter 4, including attempts to measure them directly. For now, I just want you to keep in mind that they were predicted by Einstein, and that they should act to remove energy from the binary pulsar system, as they are emitted from the orbiting bodies.

So there are three relativistic effects that can be measured in the arrival times of the pulses from the Hulse–Taylor system. These are the time delay effect from the gravitational field of the companion neutron star; the precession of the orbit of each star; and the decrease in orbital period due to the loss of energy through gravitational waves. Any two of these pieces of information can be used to infer the masses of the two neutron stars, which up until this point have not been measured in any way. The third piece of data can then be used to see if Einstein's theory is correct.

Using this method the masses of the two neutron stars in the Hulse–Taylor binary system can each be determined to be about 1.4 times that of the Sun. This is inferred using the size of the time delay effect, which can be measured to about 0.02 per cent accuracy, and the amount of precession of the orbit, which can be measured to an accuracy of about 0.0001 per cent. From the masses of the two stars, we can then calculate what Einstein's theory predicts for the amount of energy lost through gravitational waves, and what this means for the orbital period. It turns out that Einstein's theory predicts the orbits should decrease by about 3.5 metres per year. This is exactly what is measured, up to the accuracy of the observations, which is currently around 0.2 per cent. This is another spectacular confirmation of Einstein's theory of gravity, and one for which Hulse and Taylor were awarded the Nobel Prize in Physics in 1993.

One last effect that is expected to occur in the Hulse–Taylor pulsar, is *geodetic precession* (the change in direction of the axis of a spinning top). This effect will be encoded in the shape of the signals that arrive from the binary pulsar, because the pulsar itself acts as a spinning top in its orbit around its companion. Although it has probably been observed, the interpretation of this data is currently not good enough to give another precision test of Einstein's theory. This is mainly due to the unknown details of the region that emits the pulses of radiation on the surface of the neutron star.

The Hulse–Taylor binary pulsar offers some excellent tests of gravity, but while it is unique historically, it is now no longer the only binary pulsar that we know about. Let us now consider newly discovered systems that in some cases can already rival the tests of gravity we can perform with the Hulse–Taylor system, and which promise to outstrip it in the future.

Other binary pulsar systems

Due to its historical significance, the Hulse–Taylor binary pulsar has the privilege of being named after its discoverers, as well as having a scientific name (PSR B1913+16). Other binary pulsar systems are usually just referred to by their scientific names. The convention that's been adopted for this is to call the system *PSR*, for 'Pulsating Source of Radiation', and then to write its position on the sky in terms of right ascension and declination (these are coordinates that indicate positions on the sky). The letter 'B' or 'J' is also used to denote whether the pulsar was discovered before or after 1993 (the ones discovered after this date usually have their position recorded to higher accuracy).

Up until 2006, we only knew about the existence of eight other binary pulsar systems that had orbital periods of less than a day. Some of these systems have special properties that make them particularly interesting for studying gravity, and although they haven't been observed for as long as the Hulse–Taylor pulsar, they still offer new insights into how gravity works. In the rest of this section, I will give a very brief summary of some of the most interesting of these systems, before finishing off with a look forward at what the future holds for extrasolar tests of gravity.

Let's start with PSR B1534+12. As the name shows, this binary pulsar system was discovered before 1993. The remarkable thing about this system is that we see it almost exactly edge on. That is, our line of sight to the system lies almost exactly in the orbital plane of the two neutron stars. This amplifies the time delay of the radio signals, as at some points in the orbit the radio waves from the pulsar have to pass very close by to its companion before they can make their way to Earth to be observed by our astronomers. The pulsar also has particularly strong and narrow radio pulses, which makes it a very good clock. Unfortunately the distance to this system is not known with much accuracy, which limits the precision with which it can be used as a test-bed for gravity. To date, this system has therefore not provided as much information about gravity as the Hulse–Taylor pulsar.

Another binary pulsar system of special interest is PSR J1738+0333. This system is thought to contain a pulsar in orbit around a white dwarf (the bigger brother of the neutron star, described earlier in this section). The special thing about this system is that the two objects involved are very different from each other. This allows for a new test of gravity. It just so happens that in Einstein's theory the emission of gravitational waves from a binary system isn't particularly sensitive to whether or not the two objects are similar or different. Most of the possible alternatives to Einstein's theory do, however, predict sensitivity to this type of difference. By looking at how much energy is lost to gravitational waves in a system such as PSR J1738+0333, it is therefore possible to provide an additional test of Einstein's theory. If Einstein were wrong, then we should expect PSR J1738+0333 to lose energy at an anomalously high rate. So far, no such anomaly has been detected, which provides still further verification of Einstein's theory.

However, probably the most exciting system that was known about before 2006 is PSR J0737-3039A/B. This system was discovered in 2003, and has a number of almost unbelievable properties. Chief amongst these is the fact that both the neutron stars in the system were observed to be emitting pulses of radiation leading to it being called *the double pulsar*. No longer was it the case that the second object in the system was simply a passive companion, providing only a gravitational field through which the radio signals of the pulsar travelled. In this system both objects were emitting pulses of radiation, which allowed both of their orbits to be tracked in ways that were not previously possible. As if this amazing discovery wasn't enough, however, it also turned out that the two pulsars were moving at extremely high velocities (even by the standards of binary neutron stars), and that the system was almost exactly edge-on. This combination of properties greatly enhanced the effects of relativistic gravity in the system, to such an extent that by 2008 one of the pulsars had precessed so far that its radio pulses went completely out of view.

No longer do we have to observe for many decades to see the subtle effects of Einstein's gravity; with the double pulsar system we could see them in just a few short years. It is now the case that PSR J0737-3039A/B provides even better evidence for the existence of gravitational waves than does the Hulse–Taylor pulsar. While both pulsars were visible, this system provided *six* ways to measure the gravitational fields of the neutron stars compared with the three that are available for the Hulse–Taylor system. After determining the two unknown masses of the neutron stars, this leaves four independent tests of gravity in a single system. Yet again, Einstein's theory passed these tests with flying colours.

The future

While astonishing discoveries have already been made by observing gravitational systems outside of the Solar System, it is highly likely that we have even more to look forward to in the future. The reason for this optimism is partly due to the construction of a new generation of telescopes, the largest of which is known as the *SKA* (Square Kilometre Array). The SKA will be a telescope designed to receive radio waves from distant sources, and it will be built on a scale never before seen on Earth.

The SKA will consist of thousands of radio antennas and dishes, spread over distances of several thousand kilometres in South Africa and Australia as well as several other sub-Saharan states.

The total collecting area of the telescope (the sum of all dishes and antennas combined) will be one million square metres. It will be fifty times more sensitive than any other radio telescope ever built, and will require a computer network with a capacity larger than all current internet traffic combined. The estimated cost of the SKA, at the time of writing, is around €2 billion, which is being supplied by an international collaboration between Australia, New Zealand, Canada, China, India, South Africa, Italy, Sweden, the Netherlands, the UK, and Germany. By any standard, the SKA is a monumental undertaking.

A project the size of the SKA takes a long time to plan and build, but when it starts taking data in 2020 it will allow experiments to be performed that were never previously possible. For the purposes of studying gravity, one of the most important of these will be the measurement of large numbers of pulsars. First of all, the SKA is very likely to discover a large number of new binary pulsar systems, each of which can be used to study gravity in the way just described. Second, and perhaps even more exciting, the SKA will likely detect hundreds of rapidly rotating *microsecond pulsars* (i.e. pulsars that rotate millions of times a second). The idea is that by carefully measuring the arrival times of the signals from each of these pulsars, the SKA will be able to directly measure the effect of long wavelength gravitational waves as they pass through our part of the Universe. In this way, the SKA promises to be a giant gravity wave detector (something I will explain in more detail in Chapter 4).

It is expected that the observations of gravitational phenomena made by the SKA will be about a hundred times more accurate than those made by observing bodies within the Solar System. This will be a huge leap forward. At present, binary pulsar observations are just starting to overtake observations made in the Solar System as the best testing ground for gravity. The SKA will do considerably better. Beyond even this though, the SKA offers a few more tantalizing possibilities for testing gravity. One of these involves using the SKA to test how gravity works over very large distance scales in the Universe. I will return to this in Chapter 5. Another is the possibility of discovering a pulsar in close orbit around a black hole. A black hole is expected to be the most extreme gravitational field that exists anywhere in the Universe. It is the result of a large star collapsing in on itself at such a rate that nothing can stop it. The existence of binary systems comprised of a pulsar and a black hole are expected to be rare, but if they do exist, then there's a good chance the SKA will find them. Such a system would provide the opportunity to test gravity in the most extreme of all environments. It is a very exciting prospect.

Another, more immediate, reason for being hopeful about the future of extrasolar gravitational physics is the recent discovery of another new type of pulsar system. In 2014, a team of astronomers announced that they had discovered a pulsar in orbit with not one, but two, white dwarfs. They named this triple system *PSR J0337+1715*. The orbits that are possible in a triple star system are much more varied than the possibilities in a binary system, and it looks as though the system is structured in a hierarchical way, such that the pulsar is in a close orbit with one of the white dwarfs, while the second white dwarf orbits them both at a larger distance. The outer white dwarf appears to be accelerating the orbits of the inner pair, providing a new type of laboratory within which strong gravitational physics can be studied.

It's expected that new telescopes such as the SKA will aid the study of both double and triple pulsar systems. It may even be the case that it allows astronomers to observe the frame-dragging effect (described in Chapter 2), in which space itself is dragged around with the rotating stars. Such a measurement would not only be of interest for the study of gravitational physics, but also for astrophysicists who want to know how matter inside neutron stars behaves. If such observations could be made, they would allow us to study a material with a density of around a trillion kilograms per cubic centimetre.

Chapter 4
Gravitational waves

Gravitational waves have already been mentioned, in Chapter 3. Now let's consider them in a little more detail. You will recall from Chapter 1 that, in Einstein's theory, gravity is due to the curvature of space-time. Massive objects like stars and planets deform the shape of the space-time in which they exist, so that other bodies that move through it appear to have their trajectories bent. It is the mistaken interpretation of the motion of these bodies as occurring in a flat space that leads us to infer that there is a force called gravity. In fact, it is just the curvature of space-time that is at work.

The relevance of this for gravitational waves is that if a group of massive bodies are in relative motion (such as in the Solar System, or in a binary pulsar), then the curvature of the space-time in which they exist is not usually fixed in time. The curvature of the space-time is set by the massive bodies, so if the bodies are in motion, the curvature of space-time should be expected to be constantly changing. The scientific way to describe this situation is to say that, in Einstein's theory, space-time is a dynamical entity.

As an example of this, consider the supernovae that we discussed previously. Before their cores collapse, leading to catastrophic explosion, they are relatively stable objects, much like our own Sun. In this stage of their life they should therefore be expected to curve the space-time around them in the same way that the Sun does, and should therefore have a similar gravitational field. After they explode they settle down to a neutron star or a black hole, and once again return to a relatively stable state, with a gravitational field that doesn't change much with time. During the explosion, however, they eject huge amounts of mass and energy. Their gravitational field changes rapidly throughout this process, and therefore so does the curvature of the space-time around them.

Like any system that is pushed out of equilibrium and made to change rapidly, this causes disturbances in the form of waves. A more down-to-earth example of a wave is what happens when you throw a stone into a previously still pond. The water in the pond was initially in a steady state, but the stone causes a rapid change in the amount of water at one point. The water in the pond tries to return to its tranquil initial state, which results in the propagation of the disturbance, in the form of ripples that move away from the point where the stone landed. Likewise, a loud noise in a previously quiet room originates from a change in air pressure at a point (e.g. a stereo speaker). The disturbance in the air pressure propagates outwards as a pressure wave as the air tries to return to a stable state, and we perceive these pressure waves as sound.

So it is with gravity. If the curvature of space-time is pushed out of equilibrium, by the motion of mass or energy, then this disturbance travels outwards as waves. This is exactly what occurs when a star collapses and its outer envelope is ejected by the subsequent explosion. The violent process that occurs in the star causes a disturbance in the curvature of space-time, much like the stone falling into the pond causes a disturbance in the water. And again, just as in the examples just given, the disturbance propagates outwards as waves.

The speed with which waves propagate usually depends on the medium through which they travel. For example, we know that sound waves travel slightly faster in warm air than they do in cold air. The medium for gravitational waves is space-time itself, and according to Einstein's theory, they propagate at exactly the same speed as light. Just as with light, they are also expected to have a speed that is independent of the motion of the person who observes them, and that is independent of the motion of their source. They therefore propagate information at the maximum possible speed, as nothing can travel faster than light.

The effect of a gravitational wave

To try and make sense of what a gravitational wave is, it might be useful to consider what its effects should be on a group of objects as it passes through them. For example, if you were trying to describe the waves on a pond to an alien who wasn't familiar with water, you might start by suggesting that the effect of the wave is to make the lilies on the surface of the pond move up and down in a smooth, repetitive way. Let us consider the corresponding situation with gravitational waves.

Let's start by considering a uniform cloud of gas floating in space. This removes the effect of the Earth's own gravitational field, which is many times stronger than any gravitational wave that we are ever likely to see, as well as removing any disturbance the gas might feel from interacting with anything else. If the gravitational wave travels through the cloud of gas, as in Figure 8, then the principal effect of the wave is to displace the gas in the directions transverse to its direction of propagation. That is, if the wave travels from left to right, then the particles in the gas are displaced up and down, and in and out of the plane of the page.

The effect of the gravitational wave on the cloud of gas may initially make it look a little bit like the wave is being supported by the gas itself, much like the wave in the pond is supported by the water. It is, however, quite different. With the gravitational wave it is space-time that carries the wave. The effect of the gravitational wave on the gas is therefore more analogous to the effect that the

8. An illustration of a wave passing through a cloud of gas. The wave travels from left to right, and displaces the particles of gas up-and-down, and in-and-out of the plane of the page.

water wave has on a lily that sits on the surface of the pond, rather than the water in the pond itself. That is, the gravitational wave is not a wave in the gas, but rather a propagating disturbance in the space-time in which the gas exists.

In fact, what the gravitational wave is really doing in this example is changing the amount of space that exists in the directions that are transverse to its direction of propagation. This means that although the atoms in the gas might be closer together (or further apart) than they were before the wave passed through them, it is not because the atoms have moved, but because the amount of space between them has been decreased (or increased) by the wave. The gravitational wave changes the distance between objects by altering how much space there is in between them, not by moving them within a fixed space. This is only possible because of the fact that space is not fixed in Einstein's theory but is itself dynamical.

To consider the effects of gravitational waves in more detail, let us now consider a ring of particles, and think what would happen to this ring if a gravitational wave passed through it. This is illustrated in Figure 9, where in this case the gravitational wave is being imagined to be coming upwards, out of the plane of the page. The effect of the wave is only really in the directions transverse to its direction of propagation. A ring of particles in the plane of the page should therefore give us a pretty good idea of its consequences.

9. The deformation that a ring of particles would experience if a gravitational wave passed through it, in an upwards direction (i.e. coming directly out of the page). The leftmost image shows the initial configuration, and the other images (from left to right) show the ring at four subsequent instances of time.

If the particles are initially arranged in a perfect circle, and are not attached to anything or each other, then the effect of the wave will be to compress the circle in one direction and to stretch it in another. The result is that the particles start to form an ellipse. As the wave passes through, it stretches the circle smoothly until it reaches a maximum deformation, at which point it stops and then reverses the process until the direction in which it initially stretched is at a minimum. This stretching and squashing in different directions is illustrated in Figure 9, and continues until the wave has passed.

The consequences of the emission of gravitational waves has already been discussed for the binary pulsar. The waves carry energy away from the system, so that the two neutron stars slowly circle in towards each other. The measured rate of in-spiral in the binary pulsar has given good evidence for the existence of gravitational waves, but it is of great scientific interest to try and see the effects of these waves directly as well. This was done for the first time in September 2015 by the LIGO experiment. The direct detection of gravitational waves is particularly exciting because it provide us with an entirely new way to view the Universe. For the first time, we don't need to rely on light to see distant objects as we can now look at them through their gravitational fields directly. This allows us to see what happens, for example, when black holes collide.

The direct detection of gravitational waves also allows for new and exciting tests of Einstein's theory of gravity. In general, one could conceive of a number of different possible effects from a passing gravitational wave. For example, the area of the circle of particles in Figure 9 could have been changed, or there could be deformations of space along the direction of propagation of a gravitational wave (as well as in transverse directions). It is the equations that Einstein proposed for the curvature of space and time that forbids these possibilities within his theory, but if he was wrong then they could be there in nature. By using experiments like LIGO, we can look to see if gravitational waves have the specific effects that Einstein predicted. This provides yet another test of his theory. Beyond this, being able to gain a new window into what happens when black holes collide provides a number of exciting possibilities for the further study of gravitational physics.

Gravitational wave detectors

The existence of gravitational waves was first predicted by Einstein in 1916, and, after nearly a century of effort, they were detected for the first time in 2015. The reason it took so long to do this is largely a consequence of the extremely low amplitude of the signal. In the diagram shown in Figure 9, the effect of the gravitational waves was enormously exaggerated in order to try and make it easier to visualize their effect. In reality, the fluctuations in the shape of the ring should only be expected to be at the level of about one part in a hundred million trillion. That is, if we made a ring of particles that was a 1,000km across, we shouldn't expect its shape to change by more than a trillionth of a centimetre. This is obviously very difficult to detect.

Despite the difficulty, or perhaps because of it, a huge amount of effort has been expended in trying to detect gravitational waves directly. Much of the early history of this work involved what are now called *Weber bars*. These instruments, named after Joseph Weber of the University of Maryland, consisted of large cylinders of metal. They were about a metre wide, and a couple of metres long. The idea was that if a gravitational wave passed through the Earth, and hence also through the detector, then it might cause the bar to start ringing, like a bell hit by a hammer. For this to happen, the gravitational wave needs to be of just the right frequency. If such a wave were to exist, it should be expected to cause small vibrations in the bar.

The detectors that were used to see if the bar was vibrating were sensitive enough to detect changes in length of 1 part in 1,000 trillion. Remarkably sensitive as this is, it's not enough to measure the gravitational waves that we now know to exist. This didn't stop a few false alarms being sounded though. In 1968, Weber claimed he had evidence for the existence of gravitational waves, which would have made him a prime contender for a Nobel Prize. Unfortunately, the claimed detection couldn't be reproduced, and it is now widely believed that it was erroneous.

Some modern versions of the Weber bar are still being used today. An example is the MiniGRAIL experiment in Leiden. It consists of a 1,150kg metal sphere, and is about 1,000 times more sensitive than the ones used by Weber himself. However, most modern attempts at detecting gravitational waves use a different technology, known as *interferometry*. Detectors based on this technology are similar, in principle, to the Michelson–Morley device described in Chapter 2. The modern devices are, however, much, much larger. The largest of all, at the time of writing, is the LIGO detector, in the USA.

LIGO, or the Laser Interferometer Gravitational-Wave Observatory, has two bases of operation; one is in Livingston, Louisiana, and the other is in Richland, Washington. Each site contains a giant interferometer. The interferometers consist of two 'arms', which are built to be at right angles to each other, as illustrated in Figure 6. Each arm is several kilometres long, and contains a tube of near vacuum about a metre wide. Lasers are fired down each of these tubes, and are reflected from hanging masses at the end of each. When the laser light makes its way back to the point where the arms meet it can be studied, and the length of each of the arms inferred from the pattern it makes when the two laser beams interact. When a gravitational wave passes through them, the arm lengths change, and the patterns made by the interacting laser light also change.

The LIGO detector is exquisitely accurate, and has a sensitivity that is about a million times greater than a Weber bar. The technical challenges required to reach this accuracy are phenomenal. The experimentalists have had to overcome all kinds of rogue signals that pollute the data. These range from seismic noise, in which the tiny internal motions of the Earth cause the reflecting objects to vibrate, to the vibrations caused by a strong wind outside. In fact, the detectors are now becoming so precise that one of the limiting factors is currently the fact that laser light is made up from photons, so that a continuous signal from the reflecting objects is not possible.

Despite these great problems, and despite the enormous financial, political, and engineering difficulties it has taken to overcome them, the LIGO detector has ultimately been a great success. On 14 September 2015 it detected, for the first time in human history, direct evidence for the existence of gravitational waves from colliding black holes. It's almost impossible to overstate the significance of this event, which will probably go down in history as one the greatest scientific achievements of the modern age. So let's consider it in more detail.

The observation of gravitational waves by LIGO

At precisely 45 seconds past 9:50 a.m. (Greenwich Mean Time), on 14 September 2015, the LIGO gravitational wave detector at Livingston, Louisiana, detected a fluctuation in their interferometer. This signal lasted for only 0.2 seconds, and appeared to make the 4km-long arms of their interferometer change length by about 1/1,000 the width of proton. About 0.007 seconds later, a similar signal was detected by the detector in Hanford, Washington. This set alarm bells ringing, and the scientists who ran the experiment were soon in little doubt: they had detected a gravitational wave passing through the Earth.

Before we think about the event that caused the wave, let's consider the signal itself. If you were to look at the data from either LIGO site, it would appear as little more than a blip on the back of the constant noise that exists within the detectors. The signal, at its peak, was only ever about twice the amplitude of the random noise that pollutes the detectors' output. This makes it difficult to see, and difficult to confirm it as a true detection. The reasons why the scientists were so convinced it really was a gravitational wave are twofold. First, the wobble in the two detectors, at the two different locations, were very similar. It might be possible to have a random tremor at one location on the Earth that caused something that looked like a gravitational wave, but to have exactly the same tremor at almost the same time at two different geographic locations is highly unlikely. Second, and just as importantly, the LIGO scientists knew what shape the wave should take. This allowed them to scan the data, in a process known as *match filtering*, to look for their signal. Taking these two facts into account means that the LIGO scientists are more than 99.999 per cent confident that the wobbles they observed really were due to a passing gravitational wave and not to any spurious source of noise.

This result is clearly an enormous achievement for the experimental team at LIGO, who produced and operated two of the most sensitive pieces of scientific equipment that have ever existed. It is, however, also a huge achievement for a very large number of theoretical physicists. This is because the match filtering process I mentioned before, and which is crucial for the detection, requires the detailed modelling of some of the most extreme gravitational fields in nature: merging black holes. It is only by understanding the details of what happens when black holes collide that scientists were able to produce filters for the expected gravitational wave signals that LIGO might detect, and this is a lot more difficult than it sounds. There are a number of different ways that two black holes can merge, and an awful lot of mathematical and computational work has been needed to understand exactly what Einstein's theory predicts for the gravitational waves that should be emitted from such systems.

I won't go into the lengthy details of any of the mathematics, but we should consider the actual astrophysical event that led to the emission of the waves. From the shape of the signal in the detectors, and the detailed modelling of merging black holes described earlier, it is thought that the waves were created by two black holes that spiralled towards each other before eventually merging, and settling down to form one large black hole. The two black holes that came together in this scenario were around twenty-nine and thirty-six times the mass of the Sun, and they settled down to a final end state in which the resulting black hole was about sixty-two times the mass of the Sun. This might sound rather peaceful, but it's actually one of the most violent processes that can occur in nature.

The astute reader will notice that twenty-nine plus thirty-six does not equal sixty-two. This is because the gravitational waves emitted during the merger carried away 3 solar masses of energy (remember that $E = mc^2$, in Einstein's theory). This is an extraordinary amount of energy for a system to lose, and goes some way to illustrating why it should be thought of as a violent event. To give some context to this number, consider the following: while these two black holes were emitting gravitational waves, they were giving out more energy than all the stars in the observable universe put together. They were also observable from 1.3 billion light years away, which is a sizable fraction of the entire observable universe. This, then, was a very extreme event, and a highly exciting one, from the point of view of gravitational physicists.

Future prospects

Although gravitational waves have now been detected, there's no reason for us to rest on our laurels. The detection by LIGO, as well as being the end of a long quest, is also the start of a new type of astronomy. One part of the reason for this hopeful statement is that LIGO will, with any luck, continue to detect more and more of these events. Another, however, is due to the new generation of gravitational wave detectors that is being planned and built. This includes the possibility of building a LIGO site outside of the USA, with India being the current favoured location. An extra site of this type would increase the ability of LIGO to determine the direction on the sky from which a gravitational wave originated, and hence improve the prospect of using gravitational waves as a new tool for doing astronomy without light.

Beyond LIGO, another possibility for the future detection of gravitational waves is a project called *eLISA* (the Evolved Laser Interferometer Space Antenna). The eLISA mission is a European Space Agency proposal to create a detector in space. This brings certain benefits over Earth-based gravitational wave detectors, the foremost of which is that it would be immune to seismic noise. This means it would be sensitive to a range of frequencies that is extremely hard to detect from the ground. It could also be much larger than Earth-based detectors, as the lasers can simply be fired between satellites, without any protective housing. The proposal for eLISA is to create a triangle of laser beams between three different satellites, each of which is separated from the others by a few million kilometres. When it comes to gravitational wave detectors, bigger is better—so eLISA is a great prospect for future detections.

But while the arm lengths of eLISA can be huge, and the seismic noise is completely absent, it does face its own challenges. The environment in space is close to vacuum, but isn't entirely empty.

The charged material that the Sun throws off should be expected to interfere with any space-based detector of this kind, as should the cosmic rays that continuously bombard the Earth. On the ground we are shielded from this interference by the Earth's atmosphere and magnetic field, but a gravitational wave detector in space would not be. And, of course, it is much more difficult to arrange and maintain an experiment when it's in space. Nevertheless, there are high hopes that eLISA will be built, and that it will detect gravitational waves in space.

One further way in which gravitational waves might be detected in the future is using cosmological observables. Cosmology is the study of the state and evolution of the Universe as a whole, and there are considerable prospects for finding the fingerprints of gravitational waves in a number of upcoming cosmology missions. We will discuss these in more detail in Chapter 5.

Chapter 5
Cosmology

Cosmology began as a scientific discipline at the beginning of the 20th century, with the work of Albert Einstein and Edwin Hubble. It was Einstein that provided the theory within which it was possible to sensibly consider the entire Universe, and it was Hubble that provided some of the first observations that showed the Universe was expanding. Before the 20th century neither of these things had been possible, and cosmology had remained almost exclusively within the province of religion and philosophy. Since then it has flourished as a science, and it is currently in the process of becoming a precision science.

The gravitational interaction is fundamental to the study of cosmology, as gravity dominates over all other forces on large distance scales. Unfortunately, it is not possible to create consistent models of the Universe with Newton's theory of gravity alone. This is because Newton assumed that his inverse square law of gravity is applicable to everything in the Universe, and that it is transmitted instantaneously. This means that, according to Newton, the gravitational field we experience on Earth should be a sum of the gravitational fields of every object that exists in the entire Universe. This isn't necessarily a problem by itself, but it does become problematic if you try to add up the gravitational fields of infinitely many objects. In this case it turns out that Newton's theory tells us that the total gravitational field at any given point in the Universe depends on the order in which we add up the gravitational fields of all of these objects. This is obviously not a very satisfactory situation.

Of course, we now know that Newton's theory is only an approximation of the more complete theory developed by Einstein. Thankfully, the problem described above does not occur in Einstein's theory. Instead, we get a rich set of models that are self-consistent, and that can be used to model the Universe we see around us. In fact, because of Einstein's focus on space and time, we get a much deeper understanding of the Universe through his theory than we ever would have got from Newton's. This is because, using Einstein's theory, we are not only able to model how objects in the Universe move, with respect to each other, we are also able to create a model of how the space and time that make up the Universe behave. Let us now consider this in more detail.

The modern history of cosmology

Modern cosmology, as we now understand it, began in Russia in the early 1920s with the work of Alexander Friedmann. Using the recently published General Theory of Relativity, Friedmann showed that a universe that was the same at all points in space, and that looked the same in all directions, should be expected to either expand or contract. This remarkable prediction must have come as quite a surprise, because at this time no astronomer had been able to reach such a conclusion using observations. Nevertheless, Friedmann was able to produce a set of equations that such a universe would have to obey, and had even been aware that the geometry of space in these models could either be flat or have positive or negative curvature. That is, he was aware that there existed solutions to Einstein's equations in which the geometry of space could be curved, like the surface of a giant three-dimensional sphere, or the surface of saddle (see images in Figure 10).

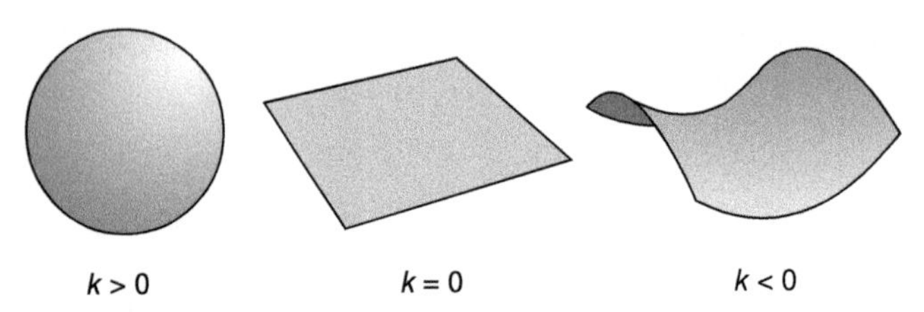

10. Examples of spaces with constant positive curvature ($k > 0$), vanishing curvature ($k = 0$), and negative curvature ($k < 0$), respectively.

Friedmann was a pioneer, but his work was not widely recognized at first. He was initially criticized by Einstein, who thought he was in error. Einstein later introduced an alternative model of the Universe, which he forced to be static by introducing a new term into his equations that he called the *cosmological constant.* This model was shown to be unstable by the Belgian priest Abbé Georges Lemaître, who was developing similar ideas to Friedmann towards the end of the 1920s. In fact, Lemaître, who had been a colleague of Sir Arthur Eddington, had written in a scientific paper in 1927 that observations suggested that the Universe was indeed expanding, in what later became known as *Hubble's law*. The article containing this monumental discovery was initially published in French, in an obscure Belgian journal. Strangely, when it was translated into English, in 1931, the section on Hubble's law was missing. Nevertheless, Lemaître is remembered today as one of the most important figures in the development of modern cosmology.

Friedmann and Lemaître were both mathematicians, and although the latter had a good knowledge of astronomy, it was not until the astronomer Edwin Hubble published his famous results in 1929 that cosmology really got started as an observational science. Hubble showed the world that the Universe was expanding. He did this by calculating the distances to what we now refer to as *galaxies*, and by using known information about their motions. Hubble showed that the recessional velocity of a galaxy is proportional to its distance from us (so that, e.g., if galaxy A is twice as far away as galaxy B then it should recede at twice the velocity). This is just what Lemaître had predicted from Einstein's theory, and it proved beyond reasonable doubt that the Universe was indeed expanding. Einstein gave up on the idea of a static universe, and described the cosmological constant as the 'biggest blunder' of his life.

The expansion of the Universe might perhaps seem like quite a different phenomenon to those we usually think of as being due to gravity, but it isn't really. The large-scale expansion of the Universe is intimately connected with gravity. In a very real sense, one can think of the expansion that Friedmann, Lemaître, and Hubble discovered as being due to nearby galaxies falling away from each other under their mutual gravitational interaction. As a more domestic example of the same phenomenon, consider throwing a tennis ball directly upwards into the sky. Normally the tennis ball will reach a maximum distance from the surface of the Earth, before it starts falling back down. In the period before this, however, when the ball is travelling upwards, it is still being acted on by gravity, and it is by using the equations that govern the gravitational force that we can calculate the properties of its motion, such as how fast it will be moving at any given time in the future. Two nearby galaxies are very similar to this. The galaxies may be moving apart, but the rate at which they move, and whether or not they will fall back towards each other, is dictated by the gravitational force between them. Einstein's theory simply allows us to construct a consistent picture of an entire Universe filled with objects that are flying away from each other.

The tennis ball analogy raises an obvious question. If galaxies are flying away from each other, like the tennis ball flies upwards from the surface of the Earth when we throw it, then does this mean the galaxies will eventually stop flying apart, and start falling back towards each other? Or in other words, could the Universe eventually stop expanding and start to re-collapse? This is a perfectly good question, and the answer can again be given by considering the tennis ball. If instead of throwing the tennis ball upwards we launch it at high speed, from some super-powered cannon, then it's possible it might never come back down to Earth. Scientists call the speed required to make this happen the *escape velocity*, and it's very easy to calculate. If the tennis ball is launched at a speed greater than the escape velocity then it will never return to Earth. If it's launched with a speed less than the escape velocity, it will eventually fall back down. The situation with galaxies is very similar. If they are receding away from each other with sufficient velocity then they should be expected to fly away from each other forever, and the Universe should then be expected to expand forever. If the rate of recession is too low, then the galaxies will eventually fall back towards each other, and the Universe will start to collapse. The rate of recession between galaxies is known as the *Hubble rate*, and the velocity required to make them fly away from each other forever is known as the *critical* rate. Theory doesn't tell us if our Universe is expanding above or below the critical rate. To discover this, we have to point our telescopes out into space, and observe.

By observing the expansion of the Universe, we therefore have another way to observe the consequences of the gravitational interaction. Indeed, in this way we can ask and answer questions about gravity that cannot be easily probed by experiments in the Solar System. These include questions such as: Has gravity always had the same strength? Does light have its own gravitational field, in the way that Einstein's theory predicts? And what happens to gravity when the density of matter becomes very large? The reason why we can answer these questions in cosmology is because of the very large length scales involved; because of the fact that the Universe is expanding; and because of the finite speed of light. Let's think about how this works.

In most situations in life we're used to the idea that we can see what's happening, as it happens. However, this isn't quite true. It just appears that way. Because light has a fixed velocity (about 300 million metres per second), it takes some time for the light emitted or reflected off an object to actually reach our eyes. The speed of light is very high, so we don't usually concern ourselves with this delay. When an object is very far away, however, the delay can become significant. If the Sun suddenly exploded, for example, it would take more than eight minutes for us to know anything about it, because that's how long it takes the light emitted from the Sun to reach us (and nothing can move faster than light). Another way of thinking about this is that when we look at the Sun we see it as it was a little over eight minutes ago. The same thing happens in cosmology, but the observable Universe is very much larger than the distance from the Earth to the Sun, so the effect becomes huge. For example, it takes more than four years for light from the nearest stars to reach us, and some tens of thousands of years for light from the nearest galaxies. If we look even further away we see objects as they were billions of years ago. In a sense, we can see back in time by looking far away, and if we look far enough away we can see what the Universe looked like when it was very young.

Now, it's a well-known result in thermodynamics that when you compress an object (like a balloon full of air), it gets hotter. Likewise, if you make the same object expand then it gets cooler. The Universe is no exception to this rule. If we think of the expanding Universe as playing on a movie reel, then if we run the reel backwards we should expect to see the Universe getting smaller and hotter, until at very early times it bursts into flames. Now recall that we can in fact see the early stages of the Universe's evolution, and you might expect that we should see a fireball if we look far enough away (and hence, look far enough back in time). This possibility was first predicted by Ralph Alpher and Robert Herman in the late 1940s, but it wasn't until 1965 that it was accidentally observed by the radio astronomers Arno Penzias and Robert Wilson. The signal they detected is now known as the *CMB*, or the Cosmic Microwave Background.

The discovery of the CMB confirmed to the world that astronomy could be used to see the very early stages of the Universe's evolution, when it was in an entirely different state of being. At the same time it opened the door to testing gravity in wildly new environments, where the gravitational field of light could be stronger than that of normal matter, and where we can consider time and distance scales that range over the entire observable Universe.

The early Universe

Since the early days of the 1960s, cosmology has blossomed into a well-studied field of both observational and theoretical physics. The positions of hundreds of thousands of galaxies have been mapped, we've seen astrophysical events that occurred many billions of years ago, and the CMB that Penzias and Wilson discovered has been measured to incredible accuracy. These observations, and others, have been used to give precise answers to questions such as 'How old is the Universe?', 'Will the Universe expand forever?', and 'What types of matter exist in the Universe?'. The answers to these questions are somewhat puzzling, but have profound consequences for our understanding of gravity. We will consider them in this section.

Let's start at the beginning of time. If the Universe was smaller and hotter in the past, then if we consider earlier and earlier times, then the density of matter should be expected to become larger and larger. Now, it turns out that not all types of matter increase in density at the same rate, as we go back in time in this way. The density of light (or *radiation*, as it's often referred to by physicists) increases at a faster rate than the density of most other types of matter. This means that at very early times the density of radiation can be even higher than the density of the electrons, neutrons, and protons that make up normal matter. In this case the gravitational field of radiation becomes the dominant influence on the expansion of the Universe.

The radiation-dominated stage of the Universe's evolution is relatively brief. It only lasts for the first few tens of thousands of years after the Big Bang. Nevertheless, it is an extremely interesting period of time, particularly for the study of gravity. One of the physical processes that takes place during the radiation-dominated period is the synthesis of the light elements (hydrogen, helium, lithium, etc.). Among the many factors that influence this process, one of the most important is the expansion rate of the Universe. Detailed calculations, and observations of the amounts of hydrogen and helium we see in the Universe around us, allow us to place tight constraints on the gravitational field produced by radiation in the very early Universe. The results of such studies are consistent with the predictions of Einstein's theory, with uncertainties at the level of only a few per cent. This is less precise than observations of gravity in the Solar System or in binary pulsar systems, but it's not bad considering it involves testing what happened billions of years ago.

As well as the synthesis of light elements, however, there are other interesting physical processes that take place during the early stages of the Universe's history. One of these is the process that eventually leads to the formation of the first astrophysical structures in the Universe. It's been known since the findings of Penzias and Wilson that the early Universe looks very close to being perfectly smooth. Very close, that is, but not exactly so. There are small ripples in the CMB radiation that these astronomers found, and these ripples are thought to be the seeds of what eventually turned into the complex network of galaxies and clusters of galaxies that we see around us today. It is gravity that was responsible for the collapse of these small ripples into galaxies, but before this occurred gravity still played a crucial role.

In the early Universe there was a battle between gravity, on the one hand, which tends to make matter clump together, and radiation, on the other, which interacts with the matter, and can cause it to spread out. Any small fluctuations that exist in the interacting soup of matter and radiation therefore start to oscillate, as they're pulled together by gravity, and pushed apart by the radiation. The period of these oscillations depend on their size in space, but are easy to calculate. This fluctuation in the density of matter continues until the Universe cools to a sufficient degree that it becomes transparent (at very early times it's opaque, like a fireball, as discussed before). At this stage the radiation can stream past the matter, and reach the telescopes of distant observers billions of years later, with very little interruption. The CMB that Penzias and Wilson discovered is made up of exactly this type of radiation, measured more than thirteen billion years after the fireball ended. The effects of the war between gravity and radiation are imprinted as tiny ripples in the CMB. These ripples contain a lot of information about the rate of the Universe's expansion; the amount of radiation in the Universe; and the ways in which radiation and other types of matter interact. They also contain information about the space through which the radiation has travelled, before we observe it on Earth. In short, the CMB is a scientific treasure chest.

Detailed observations of the CMB first began with the launch of NASA's Cosmic Background Explorer (COBE) in 1989. This satellite experiment observed the background radiation over the entire sky, and showed that the radiation was of exactly the form one would expect if it was emitted from the primordial fireball. The COBE experiment also made the first attempt at observing the small ripples we just discussed. In the end, COBE didn't have the resolution required to extract much information from these ripples, but it made a promising start. Since then, a series of balloon-based experiments have been performed. Among these were the BOOMERanG and MAXIMA experiments, which were launched in the late 1990s. The detectors in these experiments had sufficient resolution to see the largest of the ripples, and this was enough information to determine that the Universe was extremely close to the 'critical' value of expansion, right at the borderline between re-collapse and eternal expansion. For this to be compatible with the rate of expansion we see around us in the Universe today, however, it looked like something strange had to have happened. Something had to have sped up the Universe's expansion between the time of the fireball and now—and by some considerable degree.

Background radiation experiments took another leap forward at the beginning of the 21st century. In 2001, NASA launched the Wilkinson Microwave Anisotropy Probe (WMAP) into space. The WMAP experiment was able to see not only the largest of the ripples, but some of the smaller ones too. This was tremendously important, as it allowed the processes involving the growth of these ripples in the early Universe to be made the subject of observational scrutiny. This was followed, in 2009, by the launch of the Planck Surveyor, by the European Space Agency. Planck was a step beyond WMAP, and allowed many more of the ripples to be measured. The results of WMAP and Planck were a glorious confirmation of the theoretical physics that had been developed to understand the growth of the ripples in the early Universe. They showed that the collapse due to gravity expected from Einstein's theory took place just as expected, and that the amount of radiation in the early Universe was compatible with that required by the primordial nucleosynthesis calculations. They also found, however, that there appeared to be a large amount of matter in the Universe that didn't interact with radiation in any way, other than through its gravitational field. This isn't how normal matter behaves.

The background radiation contains further information beyond what I've just described. Some of this I'll describe later on, as it is more a prospect for the future rather than something that has already been detected. It's worth mentioning here, however, that as the radiation travels through the Universe, from the primordial fireball to our telescopes, it picks up a lot of information about the gravitational fields of objects in between. One way this happens is through the bending of light, as discussed in Chapter 2. The background radiation is no exception to this phenomenon, and as it passes by massive objects, its trajectory is bent by their gravitational fields. This distorts the pattern of ripples in a calculable way. It changes what the ripples look like, and is an effect that was observed by Planck. Another effect that can be seen in the background radiation is due to the evolution of gravitational fields as the Universe expands. If this happens, then a photon that enters a gravitational field with a given amplitude could find itself leaving a field with a different amplitude. The difference in these amplitudes gives (or takes away) energy from the photon. Comparing observations of this effect to the theoretical predictions gives further evidence that the expansion of the Universe is speeding up.

The expansion history

After the Universe became cool enough to see through, there was a period that astronomers refer to as the *dark ages*. This was the period after the initial fireball, but before the first stars and galaxies formed. There is very little for astronomers to look at during this part of the Universe's history, as most of the matter was in clouds of gas. After about a few hundred million years, however, the first stars and galaxies had begun to form. Since then, structures have grown continuously, and on ever larger scales, as the Universe has evolved. Of course it's gravity that has caused this to happen, and much information can be obtained about gravity by looking at the astronomical structures around us. For the moment though, let's think about how astrophysical bodies can be used to probe the expansion history of the Universe.

Hubble started this field, with his famous paper in 1929. As with most great scientific results, his work was built upon and extended by the generations that followed. The aim of all this work has been to determine how fast a distant object is moving away from us, as well as exactly how far away from us that object is. This information can then be used to determine how fast the Universe is expanding. The former of these two problems is actually relatively straightforward. Light from stars, and most other bodies, is emitted in specific frequencies that correspond to the chemical elements that it's made from. Now, when a body is in motion, as most astrophysical objects tend to be, then the frequency of the light that reaches us is shifted by the Doppler Effect. This phenomenon is the same as the shift in frequency you hear when an ambulance drives past you; the sound it emits seems higher pitched when it's travelling towards you than when it's moving away. In both cases, the change in frequency can be straightforwardly related to the velocity of the body in motion. This means that if we know the chemical elements in an astrophysical body (which we often do), then it's relatively easy to work out how fast that body is moving away from us.

Accurately determining the distance to astrophysical bodies is, however, a more challenging task. The general methodology that's used for this job is to look at objects that are reasonably nearby. If we can determine the distance to these nearby objects, which is usually a bit easier, then we can use them to calibrate the distance to similar objects that are further away. As an example of this method, let's consider the *Cepheids* that Hubble used in his famous paper. A Cepheid is a star whose brightness varies in a periodic way. It was already known that there is a relationship between the period of a Cepheid and its luminosity (its actual brightness, as opposed to its apparent brightness, which depends on distance from us). This was determined from nearby stars whose distances were known. Hubble used this information to work out the distance to faraway Cepheids. The logic is quite straightforward: you look at the Cepheid and measure its period; you use this information to work out how much light it's emitting; and you compare this to how bright the object appears on your photographic plate. There's a simple law that tells you how bright an object at a certain distance should be when it's emitting a certain amount of light, so you have all the information you need to determine its distance.

Unfortunately there are a number of things that can go wrong with this method. The rules used to determine the distance to an object (such as the relationship between period and brightness in Cepheids) might only be approximately true. You also have to assume that the distant objects observed are the same as the nearby objects that were used to determine the rule. This isn't always true, as it can be hard to identify objects when they are very distant, and because it might be that some rules change over time (recall that when you look far away, you are looking at objects as they were long ago). These problems, and more, need to be carefully considered, as they can sometimes lead to incorrect inferences. In his 1929 paper, for example, Hubble inferred an expansion rate for the Universe that's approximately ten times larger than all modern measurements. This error was due to incorrectly inferring the distances to galaxies using the Cepheids.

The current state of the art in this field is achieved using observations of *supernovae* (exploding stars), but the underlying methodology is still quite similar to the one used by Hubble. An individual supernova can be as bright as an entire galaxy, so they are reasonably easy to spot, if you know what to look for. They can also be seen from very far away. Now, there are different ways that a supernova can happen, and, of course, astronomers have given names to them all. The type of supernovae that are most useful for probing the expansion of the Universe are known as *Type Ia*. These explosions are caused by the accretion of matter on to a white dwarf from a nearby star. When enough matter has accumulated, the white dwarf can no longer hold itself up against the pressure of gravity, and it collapses and explodes. The good thing about Type Ia supernovae is that they tend to happen in a very similar way, wherever or whenever they occur. This means that, if they can be properly identified, then their brightness can be used to get quite a good estimate of their distance.

It wasn't until the late 1990s that the first results on the expansion history of the Universe from Type Ia supernovae began to emerge.

Both the Supernova Cosmology Project and the High-Z Supernova Search Team were working on this idea, and they published their first results at around the same time. Using observations of supernovae that were at vast distances, and hence that had exploded several billions of years ago, they found something very surprising. They determined that the expansion of the Universe was not slowing down, as one would expect for objects that fall away from each other under their mutual gravitational attraction, but was instead speeding up. This was entirely unexpected, and it shocked the physics community. In terms of our understanding of gravity, however, it's especially fascinating. We'll go into its consequences further in Chapter 6, but for now let's return to the large-scale structure of the Universe.

The late universe

Just as stars group together to form galaxies, galaxies group together into structures called *clusters* and *super-clusters*. These are what cosmologists are referring to when they talk about *large-scale structure*. The study of the large-scale structure of the Universe, once again, was started by Hubble. It was Hubble who realized that the spiral-shaped objects that astronomers observed through their telescopes were, in fact, distant galaxies. Up until this point it had been a real question as to whether or not our own galaxy was the only one that existed in the Universe, like an island in the infinite cosmos. Using the Cepheids that we discussed earlier, Hubble showed that the spirals were much more distant than the stars we see around us. The only explanation was that they were larger bodies, themselves made up from very many stars. This started the quest to map the structures that exist around us.

As with most branches of observational cosmology, progress in this new field was initially rather slow, and only began to pick up pace towards the end of the 20th century. One of the landmark missions in this field was the Harvard-Smithsonian CfA survey, which began in 1977 and ran until 1995. The CfA survey measured the recessional velocity of almost 20,000 galaxies, and recorded the position of each of them on the sky. Using Hubble's law, they then converted these velocities into distances, and started to map out the structure that existed in the Universe on very large scales. They discovered that galaxies clump together to form structures that span enormous distance scales. One of the most impressive of these is what is known as the *CfA2 Great Wall.* This structure is a huge concentration of galaxies, which is so large it would take light more than half a billion years to get from one end to the other.

More recent galaxy surveys have discovered even larger numbers of galaxies. The 2dF survey, which used the Anglo-Australia Telescope in New South Wales, ran from 1997 to 2002, and observed more than 200,000 galaxies. The Sloane Digital Sky Survey (SDSS), which started in the year 2000 and is planned to continue until 2020, has so far measured millions. In fact, there are now so many images of galaxies (and other astrophysical bodies) that it is impossible for astronomers to go through all of them individually. Computer programs can be used for this, but they tend to be worse at recognizing important features than the human eye (and brain). A clever way around this problem was therefore to put the images online and let the public take part in identifying them—a project known as 'Galaxy Zoo'.

Many more structures were found by 2dF and SDSS, and on even larger scales than the CfA survey. The biggest of these was the *Sloane Great Wall,* which is around twice as big as the CfA2 Great Wall. In fact, the Sloane Great Wall is so big that if you took similar sized structures and put them end to end, you could only fit a few dozen in the entire observable Universe. It's truly enormous, but one should bear in mind that this is still only a fraction of the distance probed by supernovae and the CMB. There are many more galaxies out there waiting to be discovered, and it remains to be seen if there are any structures that are even larger (the expectation is that there are not, but expectations are not always realized).

This is all very impressive, but let's return to what it means for the study of gravity. The structures that are observed in these surveys are the result of gravitational attraction. At very early times the Universe looked smooth, as verified by observations of the CMB. In order to get from that state to the present day situation, where there exists vast networks of structure, the matter in the Universe must have clumped together. The way in which this should happen is thought to be well understood on large scales, but starts to get a bit more complicated on small scales. Both of these regimes contain a wealth of information for those who are interested in gravity, so let's consider them separately.

On large scales the growth of structure happens in a predictable way. This is essentially because the large-scale bulk motion of the matter in the Universe is small on these scales when compared to the cosmological expansion. The growth of structure on large scales is, however, very sensitive to the precise rate of cosmological expansion. If the expansion is dominated by normal matter, then structures grow. This happens on smaller scales first, and on larger scales later on. Now, because we know what the seeds of structure look like, from the CMB, we can calculate what we expect the structure on large scales to look like, and we can compare this to what astronomers actually see. The results are very interesting.

First, the results of observing structures on large scales indicates quite strongly that there is matter in the Universe that does not interact with light. The reason we know this is because there is more structure on certain length scales than there would be if this were not true. That is, if all matter interacted with light, then the high levels of radiation in the early Universe should have suppressed the seeds of structure in a predictable way. What we see, however, is the level of structure that one should expect if radiation hadn't done this. The logical conclusion is that there exists matter in the Universe that does not interact with radiation, and that it was the gravitational field of this matter that served as the seeds of the structures that we see around us today. What is more, by looking at how much structure exists on different length scales, we can gain valuable information about how gravity works over very large distances.

Second, the large-scale structure in the Universe can be used as a kind of ruler, to measure the size of the Universe and how much it has expanded. This is because the initial ripples have a characteristic length. By comparing the scale of these ripples in the background radiation, to the scales that occur in the large-scale structure around us, we can therefore see in quite a direct way how much the Universe has expanded (as the former is the source of the latter). This leads to yet another surprising result. The Universe seems to have expanded more than it should have done if its expansion were dominated by the gravitational field of the matter within it. In other words, the rulers in the late Universe seem to be too big.

Let's now consider what happens on smaller distance scales, much smaller than the great walls discussed earlier. On these scales the velocities of astrophysical bodies, like stars and galaxies, are not necessarily small compared to the cosmological expansion. The analysis is therefore much more complicated, as the bodies move and interact in much more complicated ways. The current best method for studying this situation is to create huge computer simulations of very many bodies. The space that the bodies exist within is expanding, as Einstein argued that it should, but the gravitational fields of the bodies within the space are usually treated in the way that Newton specified. This is a vast extrapolation of Newton's ideas, but it is widely thought that it's a valid way to proceed. Let's now consider how gravity can be explored in this regime.

The first and most obvious thing to do here is to track the motion of the galaxies, and the form of the larger structures that they create. This is very tricky, as it's difficult to take into account all the complicated effects that can occur from the many astrophysical processes that take place in the Universe. A supernova, for example, could disrupt the growth of structure, or clouds of gas could enhance it. Nevertheless, one can try to model all such phenomena—and there has been a lot of progress in doing this since the turn of the century. What seems clear, as before, is that there appears to be a large amount of matter that we can't see directly, but whose gravitational field is required to make galaxies move and cluster in the way they do.

A second approach is to look at how galaxies, and clusters of galaxies, bend the path of light. You will recall that the Sun bends the path of starlight that passes close by it, and that this was how Eddington convinced the world that Einstein's theory was correct. The same can be done with galaxies. We can look at how the shapes of distant galaxies are distorted by the gravitational fields of those that are much closer to us—in a process known as *gravitational lensing*. This effect is often very small, and it's usually an enormous challenge to see it at all. If we look at the right galaxies, or collect enough data, however, then we can use it to determine the gravitational fields that exist in space. Once more, we find that there is more gravity than we expected there to be, from the astrophysical bodies that we can see directly. There appears to be a lot of mass, which bends light via its gravitational field, but that does not interact with the light in any other way. The exact amount of bending that occurs also potentially encodes a lot of information about the way that gravity behaves on the scales of galaxies, and clusters of galaxies.

Moving to even smaller scales, we can look at how individual galaxies behave. It has been known since the 1970s that the rate at which galaxies rotate is too high. What I mean is that if the only source of gravity in a galaxy was the visible matter within it (mostly stars and gas), then any galaxy that rotated as fast as those we see around us would tear itself apart. It would be like taking a head of dandelion seeds and rotating its stem quickly between your hands. If you rotate it fast enough then you would expect the seeds to fly off, as the bonds that hold them together are not strong enough to resist the forces that result from the rotation. The same is true with the stars in galaxies. That they do not fly apart, despite their rapid rotation, strongly suggests that the gravitational fields within them are larger than we initially suspected. Again, the logical conclusion is that there appears to be matter in galaxies that we cannot see but which contributes to the gravitational field.

The concordance model

Many of the different physical processes that occur in the Universe lead to the same surprising conclusion: the gravitational fields we infer, by looking at the Universe around us, require there to be more matter than we can see with our telescopes. Beyond this, in order for the largest structures in the Universe to have evolved into their current state, and in order for the seeds of these structures to look the way they do in the CMB, this new matter cannot be allowed to interact with light at all (or, at most, interact only very weakly). This means that not only do we not see this matter, but that it cannot be seen at all using light, because light is required to pass straight through it.

This is obviously a very strange state of affairs. The substance that gravitates in this way but cannot be seen is referred to as *dark matter*. The amount of dark matter required to explain the observations is not small. There needs to be approximately five times as much dark matter as there is ordinary matter. Most people, when they first hear this news, think something must have gone terribly wrong. That nature cannot be this strange. Yet, the evidence for the existence of dark matter comes from so many different sources that it is hard to argue with it. If the evidence only came from one place, you could try and make a case that whoever collected the data, or made the observations, might have made a mistake. It's very difficult, though, to make this argument for all of the different types of observations listed here. To make so many mistakes, and for all those mistakes to conspire to suggest the same result, seems highly unlikely. So we are led to the conclusion that most of the matter in the Universe is not of the type with which we are most familiar, but is instead some new type of matter that was previously unknown.

But this is not the end of the surprises. Not only do we require extra matter to give extra gravitational fields in order to make structures form and light bend in the way that it's observed to do, but we also have to explain why the Universe is expanding faster than we thought it should. Recall that we can think of the expansion of the Universe as objects made of matter (such as galaxies) flying away from each other under their mutual gravitational interaction. If this is true, and if gravity is always attractive, then we should expect the large-scale expansion of the Universe to be always decelerating. That is, the expansion should be getting slower over time. The results of the various astronomical observations we have discussed, however, have shown that the expansion is speeding up. The conclusion is that there must be a type of gravitational field that is *repulsive*—in other words, there seems to be a type of *anti-gravity* at work when we look at how the Universe expands. This anti-gravity is required in order to force matter apart, rather than pull it together, so that the expansion of the Universe can accelerate. This is truly astonishing. The source of this repulsive gravity is referred to by scientists as *dark energy* (not to be confused with dark matter). There needs to be around three times as much dark energy as there is dark matter, in order to make the Universe accelerate in its expansion at the current rate.

So our current overall picture of the Universe is as follows: only around 5 per cent of the energy in the Universe is in the form of normal matter; about 25 per cent is thought to be in the form of the gravitationally attractive dark matter; and the remaining 70 per cent is thought to be in the form of the gravitationally repulsive dark energy. These proportions, give or take a few percentage points here and there, seem sufficient to explain all astronomical observations that have been made to date. The total of all three of these types of energy, added together, also seems to be just the right amount to make space *flat* (rather than positively or negatively curved, like a sphere or a saddle, as illustrated in Figure 10). The flat Universe, filled with mostly dark energy and dark matter, is usually referred to as the *Concordance Model* of the Universe. Among astronomers, it is now the consensus view that this is the model of the Universe that best fits their data.

The Concordance Model, and all of the observations that have led to it, is undoubtedly a great achievement of 21st-century physics. However, it is certainly not the end of the story. Not with regard to our understanding of the history of the Universe, nor with regards to the way we understand its contents, or the gravitational fields within it. To be blunt, the Concordance Model has a number of shortcomings. First, it appears to have started off in a particularly special configuration. For space to be so close to flat, and to have the background radiation and the distribution of galaxies look so evenly spread, the early Universe needs to have been extremely close to perfectly uniform in density. Second, some of the ripples we see in the CMB appear to be larger than the distance light could have travelled since the Big Bang. Nothing should travel faster than light, in Einstein's theory, so this is genuinely puzzling. Third, we have no idea what dark matter really is. We only know that it should gravitate, and that it should not interact with light. It doesn't show up in the Standard Model of particle physics, which has a place for every other known type of particle, and it hasn't yet been seen in any particle physics experiment. Fourth, the existence of dark energy, and its repulsive gravitational field, seems to require enormous fine-tuning in order to have the effect that we see today. A bit more of it and galaxies would never have formed. A bit less and we wouldn't have ever noticed it at all.

These four problems are the focus of much attention for physicists. The first two are thought to be solved by a period of very rapid expansion in the early Universe, called *cosmic inflation*. I'll describe cosmic inflation in Chapter 6. It is hoped that the third will be solved by extending the Standard Model of particle physics, and there are a number of proposals on how this could be done. At the time of writing, it is thought that the properties of dark matter particles can be investigated directly by using the Large Hadron Collider (LHC) in Geneva. Whether or not nature will have been kind enough to allow them to fall into the range of energy levels that the LHC can probe remains to be seen. But the last of these problems is probably the most mystifying of all. The lengths that some physicists have gone to try to explain dark energy are truly extraordinary. Again, I'll describe this further in Chapter 6.

Of course, there remain a few physicists that are sceptical that dark energy and dark matter exist at all. They maintain that we need to understand in more detail how gravity works on the scale of the Universe before we can be sure that they are really there. After all, it's only through their gravitational interaction that we know about these substances at all. If we've misunderstood gravity, we may therefore have misidentified them. Future astronomical observations will be used to investigate this possibility and to further explore the properties of dark matter and dark energy.

The future of cosmology

It often turns out to be folly to try to predict the future in science, but it seems reasonably clear that the 21st century will see significant advances in cosmology. We now know a lot about the way the Universe is expanding and the way that structures in the Universe have formed, but our current knowledge will be dwarfed by observations that will happen over the next couple of decades. Much of the motivation for this work comes from dark matter and dark energy. The search for these dark materials will shed further light on gravity.

Let's start with the CMB. To date, most observations of this radiation have focused on measuring its temperature in different directions on the sky, and trying to infer what the ripples in the early Universe must have looked like. The cutting edge in this work has so far been the Planck satellite. This mission was so successful, however, that it is almost impossible to do any better with future space-based missions. What can be done, however, is build bigger telescopes on the surface of the Earth. This is currently being done in the Atacama Desert in Chile, and at the South Pole. These are two of the lowest-humidity locations on the planet, and the thin dry air makes them ideal places for looking into space. These telescopes will map the CMB to very high resolution, and will supply a wealth of information about the structures that exist in the Universe.

As well as temperature, there are other things that can be observed in the CMB data. Astronomers can also measure its *polarization* (the orientation of a set of electromagnetic waves, as illustrated in Figure 11). The polarization of the background radiation carries additional information about what happened in the early Universe, and by looking for particular patterns astronomers can infer what the gravitational field looked like very early on in the Universe's history.

Some of this information duplicates what can be deduced from the temperature, but some of it is entirely new. In particular, by looking for a characteristic curl pattern in the polarization, it is possible to deduce whether there were gravitational waves travelling around the early Universe. You will recall, from Chapter 4, that considerable effort has been expended to directly detect the gravitational waves that travel through the Earth. The polarization

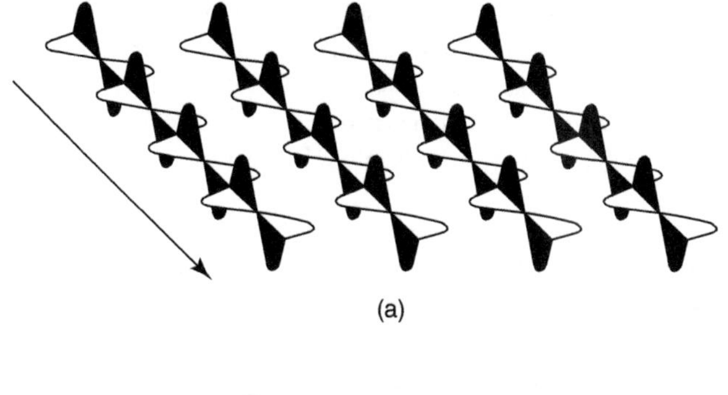

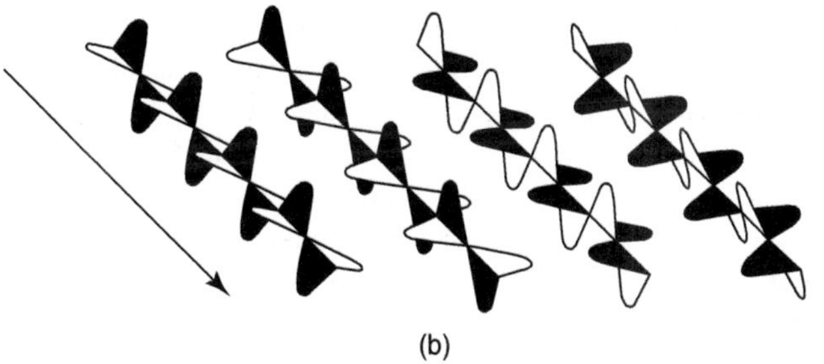

11. An illustration of (a) polarized light; and (b) unpolarized light. The orientation of the waves in the unpolarized light is random, while the waves in the polarized case are aligned. Arrows denote the direction in which the light is travelling.

of the CMB provides the scope for similar experiments in an entirely different environment.

In March 2014, scientists working on the BICEP2 experiment at the South Pole announced that they had used this method to discover gravitational waves in the early Universe. At the time of writing, it appears that the excitement generated by this announcement was premature. While the scientists had seen the curl pattern in the polarization of the CMB, it now seems that this was generated by sources closer to us and not from gravitational waves at all. This doesn't mean that gravitational waves aren't out there, in the early Universe. Future experiments will measure the polarization of the background radiation to even higher accuracy, and in more frequency bands. If gravitational waves were present in the early Universe, to any considerable level, then it's likely we'll know about them in the next decade or so. The successor to BICEP2 has already been built, and will give its first scientific results soon.

Also of great promise are the next generation of galaxy survey missions. We discussed the 2dF and SDSS surveys earlier, which were very ambitious attempts to record the positions of the galaxies in the Universe we see around us. Future surveys will be much, much bigger. Three of the grandest of these will be the Large Synoptic Survey Telescope (LSST), which is already under construction in Chile; the Square Kilometre Array (SKA), which will begin construction in 2018; and the European Space Agency's Euclid satellite, which is expected to launch around 2020. These missions will measure billions of astronomical sources, which will be used to construct maps of the Universe on unprecedented scales.

The LSST, the SKA, and the Euclid satellite will turn cosmology into a precision science. When they are in operation, we will know much more about the nature of the dark matter and dark energy: through the effects they have on the structure in the Universe; from the way in which light travels to us from those structures; and from the way that they evolve in time. In fact, it is expected that this information will be so precise that, for the first time, we will be able to start performing tests of gravity that will rival those we can already perform in the Solar System and in binary pulsar systems. This will open up a whole new window on gravity, and will allow us to test it in new ways and on new distance scales.

Chapter 6
Frontiers of gravitational physics

Throughout most of this book we have been considering the cutting edge of experimental gravitational physics research, ranging from tests on the scale of millimetres to those on the size of the entire observable Universe. In this chapter we will consider some of the issues involved in the theoretical description of gravity.

Since 1915, it has been Einstein's theory that has shaped our understanding of the gravitational interaction. This theory treats space and time as a single object, and lets the properties of that object be determined by the matter that exists within it. I hope that the reader has, by now, been convinced of the phenomenal success of Einstein's theory. It is truly extraordinary that a single theory should be able to explain such a wide array of physical effects. Einstein's theory, however, is unlikely to be the final word in our understanding of gravity. A lot has happened in the world of theoretical physics since 1915, and much of it suggests that we should expect there to be an even more fundamental theory.

Quantum mechanics and gravity

Not long after Einstein published his theory of gravity, at the beginning of the 20th century, the world of theoretical physics was forever changed (once again) by scientists such as Bohr,

Heisenberg, Schrödinger, and others. Since Newton, and up until this point, physics had been thought to be deterministic. That is, if you know enough information about the position and motions of all objects in the Universe, then you should be able to predict the future with arbitrarily high precision. Physical theories that work in this way are now referred to as *classical theories*. Einstein's relativity theory is an example of a classical theory. The revolution led by Bohr, Heisenberg, and Schrödinger created another type of theory, *quantum mechanics*. The new quantum theory was based on probability, and resulted in a description of nature in which it was only ever possible to calculate the odds that certain events will happen in the future, and within which it is never certain what the future holds.

Quantum mechanics was an astonishing success. It described the nature of light, and the building blocks of all known matter, to extraordinary precision. Later, Paul Dirac, a professor at the University of Cambridge, showed the world how these new ideas could be used to create quantum theories of electricity and magnetism. This led to what is now called the *Standard Model* of particle physics: a quantum mechanical description of all of the known particles in nature, and the forces between them. Winding forward to the present day there can be no serious doubt that nature is quantum mechanical. The predictions of the Standard Model have all been verified, with the crowning jewel being the experimental verification in 2012 of the existence of the *Higgs boson* (a theorized particle that is an important component of the modern Standard Model, and which gives the other particles their mass). It is quantum mechanics that underpins much of modern chemistry and material science, giving us the semi-conductors that our computers are made from; and the lasers and LEDs that are used to construct our DVD players and televisions. Quantum mechanics is a fact of life. It is how nature works, and it becomes increasingly important when we seek to describe the microscopic world.

Yet despite its ubiquity in all other areas of physics, the application of this approach to gravity is still not understood. While the electromagnetic force can be quantized in a relatively straightforward way, and while the matter in the Universe has long been described using quantum mechanics, the quantum theory of gravity remains elusive. This is probably the single biggest unsolved problem in physics, and has been for the past fifty years. It is extraordinarily difficult to apply the logic that has been so successful in other areas of physics to gravity, and so our current best description of gravity is still Einstein's classical theory.

Part of the problem with this state of affairs is that there are situations in which one would expect both quantum mechanics and gravity to be required. One example of this is the centre of a black hole. As we've discussed already, black holes form when large stars undergo catastrophic collapse. The matter that was originally part of the star gets compressed by gravity to ever higher densities. In fact, according to Einstein's theory, the collapse should continue until all the matter is squashed to a single point. Now, according to quantum theory we should expect new quantum effects to become apparent when we consider very small distance scales and very high energies. What I've described is therefore a regime where both quantum mechanics and gravity are expected to be required in order to get a description of the physics at play. But there is no agreed-upon theory of quantum gravity, so it is (at present) impossible to know what should happen at the centre of a star after it has collapsed. This is obviously an unsatisfactory state of affairs. If we want to be able to describe everything that exists in nature, then we need a quantum theory of gravity.

The reasons for the apparent incompatibility of quantum theory and gravity are many, and can be somewhat complicated. First there is a conceptual difference between Einstein's approach to gravity, and how forces are treated in quantum mechanics. In

Einstein's theory, gravity is a result of the curvature of space-time. There is no external force that pulls things together. The apparent way in which massive bodies move towards each other is simply a result of the curvature of space-time. The Earth, for example, is not pulled towards the Sun, it is simply in free fall, following the shortest path available to it in a curved space-time. This is not the case for other forces. The electric force, for example, is a result of the electric field that gets generated by charged particles. The electric field exists within space and time, but is not the same as space or time, in any sense. Space and time are simply the arena within which the electric force plays out. This idea of space and time having their own independent existence, and being passive parts of the problem, is built into most approaches to quantum mechanics. To try to use these approaches to describe gravity therefore goes against what we were taught by Einstein.

There are also very good mathematical reasons why gravity and quantum mechanics are incompatible. The foremost of these is a property of gravity called *non-renormalizability*. When quantum mechanics is used to describe a force, the result of the calculations that we perform can often result in answers that contain infinities. An example of this occurs when we consider two charged particles. To work out the force between them, quantum mechanically, we have to add the contributions from all of the possible positions of the two particles. Some of these configurations are when the particles become very, very close, in which case the force becomes very, very large. Summing over all possible positions then gives a result that is infinite. This unreasonable answer can be corrected for, in the case of the electric force, by using a process called *renormalization*. This process removes the parts of the equations that contributed the infinities. That is, the infinities are essentially subtracted from the original equation. The result is then a sensible answer, which can be tested with experiments. Renormalization, however, doesn't work with Einstein's gravity. The infinities cannot be subtracted, as there is nothing in the original equations that looks anything like the terms that become infinite. Quantum gravity calculations therefore seem to give infinite answers. Something is obviously wrong.

There have been many attempts to fix these problems. They range from changing Einstein's equations (so that they appear to be more renormalizable); to changing quantum mechanics (so that it is no longer based on particles); through to changing the nature of what we think of as space and time (so that they are not continuous). It is not possible to give a fair representation of all of these approaches here, or to go into any one of them in any detail. They are very complicated ideas and they are all works-in-progress. There are a couple of theories that I feel I should mention though—these are *String Theory* and *Loop Quantum Gravity*. These are both extremely bold and ambitious attempts to construct quantum theories of gravity. If correct, the hope is that physicists could use these theories to describe what happens at the centre of black holes. They are, however, quite different from each other. They prioritize different aspects of the problem, and approach the technical and conceptual difficulties just described in quite different ways.

String Theory was born out of particle physics. It is based on the idea that the basic constituents of matter are not point-like particles, but are instead tiny one-dimensional strings. This is a radical idea, and it has led to a lot of interesting maths and physics. Indeed, many physicists consider it to be our best hope of finding a theory of quantum gravity. The hypothesized strings are very small, so that for the most part they appear to us effectively as point-like particles. When we try and quantize them, however, their stringy-nature leads to different results. The equations that govern the strings also contain aspects that look a lot like the equations that govern Einstein's theory of gravity. It therefore appears that gravity is, to some extent, built into String Theory. There are drawbacks, however, as the consistency of the equations that govern the strings require us to add between six and twenty-two extra dimensions of space to our description of the Universe.

These extra dimensions are thought to be wound up and compact, so that we don't see them in our everyday lives. Nevertheless, they have to be there for the theory to be self-consistent. Interestingly, the existence of these small extra dimensions leads to the possibility that gravity could work differently on very small scales.

Loop Quantum Gravity is often seen as the main competitor to String Theory. The starting point for this theory is the idea that on very small scales space-time has a granular structure. That is, space and time are not the smooth continuous variables that we usually consider them to be. Instead, space-time is atomized. Quantum theory is then applied to the loops that make up this new structure. This is also a radical idea, and it tends to be favoured by aficionados of General Relativity because of its emphasis on space-time as the fundamental object of interest—instead of as a background. Loop Quantum Gravity is, however, very much a work in progress. It remains to be seen whether the approach it employs is the correct one or whether String Theory, or some other as yet undiscovered theory, is a preferable description of nature. Further work is needed before most of us would be prepared to make any bets on the outcome of this debate.

Particles in gravitational fields

Quantizing gravity is fraught with difficulties. So let's turn to another question: how does quantum mechanics work in a gravitational field? Here we will treat space-time in the same way that Einstein did in his classical theory of gravity. Once we have a classical space-time at our disposal, however, we will consider what happens when we try to treat the matter content as being governed by quantum theory. This mixed approach, with matter being treated quantum mechanically, and gravity being treated classically, is usually referred to as *semi-classical* physics. It is a less ambitious project than full quantum gravity but nevertheless gives us interesting insights into how quantum systems work in the presence of gravity.

One of the pioneers of this subject was Stephen Hawking, who in 1974 showed that quantum mechanics should lead to black holes emitting radiation. This discovery shocked the scientific community, as according to Einstein's theory nothing can ever escape from a black hole. Hawking's calculation was a semi-classical one. He took the classical description for the space-time around a black hole and allowed quantum mechanical particles to exist within it. He showed, using a fairly simple quantum mechanical calculation, that if there was no radiation in the distant past, then there must exist radiation in the future. The only possible explanation for this was that the radiation was produced by the black hole. Of course, radiation carries energy, and in this situation the only source of energy is the mass within the black hole (remember that mass is a type of energy in Einstein's theory). So Hawking had shown that black holes naturally shrink, by radiating away their mass, and that they must eventually cease to be.

Hawking's result was novel, and sparked several new fields of research in gravitational physics. Almost immediately after Hawking's discovery it was shown, by Bill Unruh, that the very existence of particles can be questioned when we consider them within relativity theory. Particle physics is very much the domain of quantum mechanics, and Unruh showed that if observers are in relative motion, with one accelerating with respect to the other, then it is entirely possible for one of them to detect the existence of quantum particles while the other detects nothing at all. That is, whether or not particles exist depends on the motion of the person who is trying to measure them.

To get across the strangeness of this result, let me illustrate it with an example. Consider that you're an astronaut, floating around freely in outer space somewhere. You see nothing anywhere near you. If you then start accelerating, by holding on to a passing spaceship, say, then suddenly what you thought was empty space bursts into a sea of particles. I'm exaggerating a little here, of course. You would need to accelerate very quickly indeed to see a large number of particles. Nevertheless, the principle is sound. When you accelerate you detect particles where previously there were none. Now introduce gravity and the situation becomes even more complicated. Gravity is caused by acceleration, so by sitting at my desk, in the gravitational field of the Earth, I am being exposed to a small number of particles that wouldn't be there if I were falling freely. The number is too small to measure, but if I moved my desk so that it was near a black hole (where gravity is much stronger) then it would be an entirely different story. I would be bombarded with high-energy particles and radiation.

All of this has interesting consequences for black holes, which can now have a temperature associated with them based on the temperature of the particles and radiation that they emit. But it also has consequences for other areas of gravitational physics, including cosmology. In some ways the gravitational field of a cosmological model of the Universe is similar to the gravitational field of a black hole, and indeed Gary Gibbons and Stephen Hawking showed that the radiation that black holes produce should also be produced by the expansion of the Universe. The faster the expansion, the more radiation there should be, and the higher its temperature. This radiation isn't emitted from anything within the Universe, but is a by-product of the expansion itself. It is a direct consequence of considering quantum particles existing in a gravitational field.

Cosmic inflation

To date, the most successful application of quantum theory to gravity has probably been in the very early stages of the Universe's history. The name that physicists use to describe what happened during this period is *cosmic inflation*. In Chapter 5, we considered the Big Bang model of the Universe, and its various successes. As well as explaining a lot of astronomical data, however, the Big Bang model presents us with a few problems. One of the gravest of these is that some of the ripples we see in the CMB appear to be so large that light could not have made it from one of their edges to the other within the lifetime of the Universe. This is a very serious problem, because nothing can travel faster than light. So what could possibly have caused these ripples?

The answer is not obvious, but one possible explanation for their existence is that the Universe might have expanded very quickly in its very early history. If this happened, then very small ripples would have been forced to grow into large ones, and the problem would be solved. The hypothesized period of rapid expansion is what is called cosmic inflation. Now, as scientists, the way to test a hypothesis of this type is to try and predict other consequences that it might have, and to get out our telescopes to see if we can verify those predictions. This is difficult when considering inflation, as we do not know exactly what caused it. It also happened a very long time ago. Nevertheless, there are a number of generic predictions one can make, and that can be verified by observing the night sky. One of these is that the geometry of space should be close to flat. This matches observations, as we have seen. Probably the most impressive prediction, however, involves the application of the semi-classical physics just described.

You will recall that Gibbons and Hawking demonstrated that an expanding space creates a sea of radiation. It turns out that the radiation produced is not perfectly uniform at every point in space. The statistical nature of quantum mechanics means that random fluctuations are introduced, so that at some points there is a little more radiation, and at other points there is a little less. It's impossible to predict where any one of these quantum mechanical fluctuations will occur, but the theory does let us predict how often we should expect a randomly selected point to be over-dense or under-dense. It also tells us how we should expect the over-dense and under-dense regions to be distributed, on average. These are all predictions of semi-classical physics, and we can test the theory by looking for their consequences. In particular, we can test the idea of cosmic inflation by looking for the consequences of the quantum fluctuations that it would produce.

Now recall that in Chapter 5 we discussed the ripples that exist in the CMB. These ripples are a very important source of information for cosmologists, but so far we haven't spelt out where they came from. That is, we haven't said what caused the small seeds from which they grew. These seeds need to have a very special form in order to explain the statistical properties of the ripples that astronomers measure in the CMB, and, before inflation was introduced, there was no clear idea about where they should have come from. If inflation really did happen in the very early Universe, then one way of sowing these seeds could have been through the small quantum mechanical fluctuations that Gibbons and Hawking predicted. It turns out the ripples that were measured by COBE, WMAP, and the Planck Surveyor are just what we should expect from such a scenario.

This was a remarkable discovery. Not only has the most generic prediction of the inflationary epoch been verified, but it also appears that we have verified the peculiar calculations that result from considering quantum mechanical processes in a gravitational field. The type of radiation that Hawking had predicted in 1974 still hasn't been seen directly, but its consequences appear quite plainly in the CMB. The evidence is all there, in the maps of the CMB that have been recorded by astronomers. But, again, this is not the end of the story; there is very likely more evidence out there, waiting to be collected. The same quantum mechanical processes that generate ripples in the CMB should also be expected to generate gravitational waves. These are exactly the gravitational waves that the observers running the BICEP2 experiment mistakenly thought that they had detected in March 2014 (see Chapter 5). At the time of writing, the gravitational waves from cosmic inflation have still not yet been confirmed observationally, but if they can be found by future experiments then they will open up a whole new window on the early Universe.

The cosmological constant

We noted earlier that the acceleration of cosmic expansion is said to be caused by dark energy, but we didn't go into any detail about what dark energy might be. The truth is that we don't yet know what dark energy is, but we do have a favourite candidate: the *cosmological constant*. In this section we will consider the cosmological constant in more detail.

The cosmological constant was first introduced by Albert Einstein in 1917. At that time, it was not known that the Universe was expanding, and Einstein introduced his cosmological constant in order to produce a cosmological model that was static (neither expanding nor contracting). Of course, we now know that there is very good evidence to support the idea that the Universe is expanding. When Einstein became aware of this he withdrew his cosmological constant, which was promptly brushed under the carpet as something of a scientific embarrassment. However, the cosmological constant remained a perfectly consistent modification that one could make to the field equations of his theory of gravity. It's just that there was no need for it. Not, that is, until it was noticed that the expansion of the Universe was accelerating.

A cosmological constant can be thought of as a universal gravitational force, pulling or pushing all particles in the Universe together at the same rate. This is exactly the sort of thing that's required to make the Universe accelerate. All we have to do is make sure the cosmological constant is set up to push things apart, give it the right magnitude, and it will make the expansion of the Universe accelerate. In fact, it's by far the simplest way to make the Universe accelerate.

The cosmological constant, tuned to have the correct magnitude, fits all of the current observations. Of course, we expect the quality of this data to improve considerably over the coming decades. When this happens we will be able to see whether the cosmological constant remains a good fit—or not. If it is, then this will be good evidence for its existence. If it's not, then we will have to be more imaginative. For now, we can speculate on what it would mean if there really was a cosmological constant in our Universe. This is interesting because the cosmological constant, although simple, brings with it a number of problems.

The first and foremost problem associated with the cosmological constant is that, if it is to cause the expansion of the Universe to accelerate at the present time, then it must have been very finely tuned in the early Universe. Fine-tuning is one of the bugbears of theoretical physics. It's one thing to come up with an explanation for a physical effect, but if your explanation requires things to be arranged in an extremely special way then it starts to look less and less compelling. The fine-tuning associated with the cosmological constant comes from the fact that its value doesn't change in time (it's a constant). This means that if we want it to have the correct magnitude today, then in the very early Universe we have to pick a value that is very, very, very small compared to the energy scale of matter at the time, but not quite zero. If the cosmological constant were too large, it would have caused the expansion of the Universe to accelerate at much earlier times. If this happened, then stars and galaxies would never have formed, and there would have been no life anywhere. If it were any smaller, it would not cause the required amount of acceleration, and we would never have noticed it. To fit into this sweet spot we need to pick the magnitude of the cosmological constant to have a very particular value, with very high precision. This precision is widely thought to be at the level of about one part in 10^{120} (that's one followed by one hundred and twenty zeroes).

The cosmological constant problem, just described, is exacerbated by the fact that quantum mechanical effects also contribute to its magnitude. Given our current understanding of quantum mechanics, we would expect these contributions to throw the value way off from the very special value that is needed observationally. One might counter this argument by saying that we do not yet fully understand the quantum processes that would cause these effects, and how they work in the presence of gravity. One could speculate that there might be some reason why the various quantum contributions should cancel each other out, and that we just don't know about it yet. This might be possible, but there's a further problem. The particular quantum contributions that we expect the cosmological constant to receive are not always the same—they change during the different epochs of the Universe's expansion history. To consider the possibility that a set of quantum corrections might all cancel each other out is one thing. To assume this should happen over and over again is quite another. The fact that the cosmological constant is so very finely tuned therefore looks even more surprising when we take quantum mechanical effects into account. This is why the cosmological constant problem has been called, by some, the worst fine-tuning problem that has ever occurred in physics.

The multiverse

The problem of the cosmological constant is considered so great, and so pressing, that many physicists have started to entertain some quite drastic proposals in order to try and explain it. One of the most fantastical, and most widely considered, ideas is the possibility that there is more than one universe. If this were the case, and if the cosmological constant were somehow to take a different value in each universe, then it might be possible for us to find ourselves in a universe with any given value for the cosmological constant. Even if the value we observe looks fine-tuned, this might just mean that we are in a relatively rare universe, and that there are many other universes in which the cosmological constant takes a more natural-looking value.

This idea of many universes, or a *multiverse*, doesn't by itself alleviate the problems associated with the improbability of measuring the cosmological constant to have the value that we observe. Instead of fine-tuning the value of this constant, we are instead forced to carefully select an unlikely universe for ourselves to live in. However, if we couple the idea of a multiverse with what is known as the *anthropic principle*, then things become very different. The anthropic principle, roughly stated, says that we (as life forms) can only ever observe a universe that is capable of supporting life. This sounds obvious, but it provides a mechanism to select which of the possible universes we might be able to find ourselves within. If a particular universe contains a cosmological constant that is so large that stars and planets can never form, then it is unlikely that we would find ourselves living there. This automatically de-selects a large part of the multiverse, and makes our universe look a lot more likely.

This idea raises a lot of questions. Where are these other universes? How are they connected to ours? How does the value of the cosmological constant change between them? And how likely are we to find ourselves in any one of them? These are very fundamental questions, and although there are mechanisms for generating many universes from some theories of cosmic inflation, it is pushing the boundaries of science to say that we can investigate them as if they were physical realities. For some, the idea of a multiverse is a glorious one, motivated by observations of how gravity works on astronomical scales, and fleshed out by our theories of what happened near the Big Bang. For others it is worse than the problem it was intended to solve. Many in this latter group consider it wrong to invoke unobservable regions of space and time in order to solve a problem in our own observable Universe. While it might be self-consistent, and even well motivated, to do so, the existence of these other universes cannot be tested directly. Some people within this group therefore argue that such an approach is essentially unscientific, and belongs to the realm of metaphysics.

Whether or not the limits of science can be stretched to include a multiverse is a topic of lively debate, with different groups passionately arguing their various cases. Future astronomical missions will advance this argument by measuring the properties of whatever it is that's causing the Universe to accelerate in its expansion. The future development of theoretical physics may also shed light on the naturalness of the cosmological constant we appear to measure. For now, however, we must wait.

Epilogue

Our understanding of gravity has developed rapidly over the past century. This began with Einstein's revolutionary new theory and continued with sustained developments in our understanding of both the mathematics and the observations that can be used to probe it. I've outlined how new gravitational effects have been predicted and observed in the Solar System; in exotic astrophysical systems; and in the Universe as a whole. While I've attempted to give some idea of the elegance of the concepts involved, and the details of the wondrous physics that results, this book must inevitably be left incomplete. It is only a *very short introduction* to the subject. To understand the full profundity of what Einstein achieved, and the beauty of the theory that resulted, there is no alternative but to delve deeper into the maths and physics. This should be a rewarding experience, as Einstein's theory of gravity lets us understand space and time as they really are. It lets us imagine universes that never were, as well comprehend the one in which we live. It lets us calculate what happens in environments so alien and exotic that our everyday understanding of reality is turned entirely on its head. Yet it is almost certainly incomplete. The final words on gravity have yet to be written.

Further reading

Chapter 1: From Newton to Einstein

Russell Stannard, *Relativity: A Very Short Introduction* (Oxford University Press, 2008).

*Charles W. Misner, Kip S. Thorne, and John Archibald Wheeler, *Gravitation* (W. H. Freeman and Company, 1973).

Chapter 2: Gravity in the Solar System

Clifford M. Will, *Was Einstein Right? Putting General Relativity to the Test* (Basic Books, 1993).

*Clifford M. Will, *Theory and Experiment in Gravitational Physics* (Cambridge University Press, 1993).

Chapter 3: Extrasolar tests of gravity

Katherine Blundell, *Black Holes: A Very Short Introduction* (Oxford University Press, 2015).

*Ingrid H. Stairs, Testing General Relativity with Pulsar Timing, *Living Reviews in Relativity* 6/5 (2003) at: <http://www.livingreviews.org/lrr-2003-5>.

Chapter 4: Gravitational waves

Harry Collins, *Gravity's Shadow: The Search for Gravitational Waves* (University of Chicago Press, 2004).

*B. S. Sathyaprakash and Bernard F. Schutz, Physics, Astrophysics and Cosmology with Gravitational Waves, *Living Reviews in Relativity* 12/2 (2009) at: <http://www.livingreviews.org/lrr-2009-2>.

Chapter 5: Cosmology

Peter Coles, *Cosmology: A Very Short Introduction* (Oxford University Press, 2001).
*Scott Dodelson, *Modern Cosmology* (Academic Press, 2003).

Chapter 6: Frontiers of gravitational physics

Brian Greene, *The Elegant Universe* (Vintage, 1999).
Leonard Susskind, *The Cosmic Landscape* (Back Bay Books, 2005).

*References labelled with an asterisk are advanced-level texts, intended for readers who want to learn about the subject in all its glory. They require degree-level knowledge of mathematics and physics. Readers wishing to take part in the Galaxy Zoo project should visit <www.galaxyzoo.org>.

Index

H

I

K

L

M

N

O

P

Q

R

S

T

U

W